STATISTICAL
PROBLEM
SOLVING

QUALITY AND RELIABILITY

A Series Edited by

EDWARD G. SCHILLING
Coordinating Editor
Center for Quality and Applied Statistics
Rochester Institute of Technology
Rochester, New York

W. GROVER BARNARD	**RICHARD S. BINGHAM, JR.**
Associate Editor for	*Associate Editor for*
Human Factors	*Quality Management*
Vita Mix Corporation	Consultant
Cleveland, Ohio	Brooksville, Florida
LARRY RABINOWITZ	**THOMAS WITT**
Associate Editor for	*Associate Editor for*
Statistical Methods	*Statistical Quality Control*
College of William and Mary	Rochester Institute of Technology
Williamsburg, Virginia	Rochester, New York

ADDITIONAL VOLUMES IN PREPARATION

STATISTICAL PROBLEM SOLVING

WENDELL E. CARR

South Burlington, Vermont

ASQC Quality Press

Marcel Dekker, Inc.

Milwaukee

New York • Basel • Hong Kong

Library of Congress Cataloging-in-Publication Data

Carr, Wendell E.
Statistical problem solving / Wendell E. Carr.
p.   cm.—(Quality and reliability ; 33)
Includes bibliographical references and index.
ISBN 0-8247-8704-8 (alk. paper)
1. Quality control—Statistical methods.   2. Statistics.
3. Problem solving.   I. Title.   II. Series.
TS156.C364   1992   92-6860
519.5'02462—dc20   CIP

This book is printed on acid-free paper.

ASQC Quality Press
611 East Wisconsin Avenue, Milwaukee, Wisconsin 53201

Marcel Dekker, Inc.
270 Madison Avenue, New York, New York 10016

Current printing (last digit):
10 9 8 7 6 5 4 3 2 1

PRINTED IN THE UNITED STATES OF AMERICA

ABOUT THE SERIES

The genesis of modern methods of quality and reliability will be found in a simple memo dated May 16, 1924, in which Walter A. Shewhart proposed the control chart for the analysis of inspection data. This led to a broadening of the concept of inspection from emphasis on detection and correction of defective material to control of quality through analysis and prevention of quality problems. Subsequent concern for product performance in the hands of the user stimulated development of the systems and techniques of reliability. Emphasis on the consumer as the ultimate judge of quality serves as the catalyst to bring about the integration of the methodology of quality with that of reliability. Thus, the innovations that came out of the control chart spawned a philosophy of control of quality and reliability that has come to include not only the methodology of the statistical sciences and engineering, but also the use of appropriate management methods together with various motivational procedures in a concerted effort dedicated to quality improvement.

This series is intended to provide a vehicle to foster interaction of the elements of the modern approach to quality, including statistical applications, quality and reliability engineering, management, and motivational aspects. It is a forum in which the subject matter of these various areas can be brought together to allow for effective integration of appropriate techniques. This will promote the true benefit of each, which can be achieved only through their interaction. In this sense, the whole of quality and reliability is greater than the sum of its parts, as each element augments the others.

The contributors to this series have been encouraged to discuss fundamental concepts as well as methodology, technology, and procedures at the leading edge of the discipline. Thus, new concepts are placed in proper perspective in these evolving disciplines. The series is intended for those in manufacturing, engineering, and marketing and management, as well as the consuming public, all of whom have an interest and stake in the improvement and maintenance of quality and reliability in the products and services that are the lifeblood of the economic system.

The modern approach to quality and reliability concerns excellence: excellence when the product is designed, excellence when the product is made, excellence as the product is used, and excellence throughout its lifetime. But excellence does not result without effort, and products and services of superior quality and reliability require an appropriate combination of statistical, engineering, management, and motivational effort. This effort can be directed for maximum benefit only in light of timely knowledge of approaches and methods that have been developed and are available in these areas of expertise. Within the volumes of this series, the reader will find the means to create, control, correct, and improve quality and reliability in ways that are cost effective, that enhance productivity, and that create a motivational atmosphere that is harmonious and constructive. It is dedicated to that end and to the readers whose study of quality and reliability will lead to greater understanding of their products, their processes, their workplaces, and themselves.

Edward G. Schilling

PREFACE

These problems are presented in the same spirit as other books of recreational math problems, such as Boris A. Kordemsky's *The Moscow Puzzles* and H. E. Dudeney's *Amusements in Mathematics*. The difference is that the first 210 of the 251 problems here deal with statistics, which may be a new subject for you. It's a branch of mathematics dealing with sampling and probability problems.

This book is a fun way to learn about the fascinating variety of problems that can be handled with statistics. Learn how to go up against the casino with a high chance of walking out a winner (Problems 69 and 70), or how to design an experiment to pick the better of two manufacturing methods (Problems 180–182). The table of contents shows the categories of problems, and each problem has a short title that hopefully will catch your interest. The emphasis is on industrial applications, but the principles can be applied in any field.

The Solutions give the reasoning and formulas necessary to solve the problems. Where the reader may want more information, mention is made of one of the references listed at the end of the book. You may want one of the books beside you as you do the problems. They're all good, but if I had to pick one, it would be Acheson J. Duncan's *Quality Control and Industrial Statistics*.

Most of the solutions require at most a hand calculator. However, for more lengthy calculations, reference is sometimes made to APL. This stands for "A Programming Language," an IBM computing system. Any other computing system can be used, of course, if it does the job.

The problems first appeared in the "Problem Corner" column within *Statishare*, the monthly IBM statistical newsletter that goes to all IBM locations worldwide. Each column presents three problems, with solutions elsewhere in the same issue. The industrial problems are from my experiences during 25 years as a consulting statistician within IBM. The others are either real-life problems outside industry, or just for fun.

I hope you enjoy solving the problems as much as I did writing them!

Wendell E. Carr

CONTENTS

COMBINATIONS AND PERMUTATIONS   11

GAMES 17

CORRELATION   55

RELIABILITY   57

MOVING AVERAGES 61

SIGNIFICANCE TESTS 63

CONFIDENCE INTERVALS 71

NONSTATISTICAL PROBLEMS   77

PROBLEMS

PROBABILITY

1. TV Game Show

On a TV game show, the host secretly puts $100 in one of three empty boxes and places them on a table. After a contestant picks up a box, the host always opens one of the boxes on the table and shows that it is empty. He then offers to let the contestant swap his box for the closed one on the table. Should he swap? Find the probabilities of winning the $100 with the closed box on the table, and with the box in hand.

2. Chemical Spill

For a certain chemical transfer in a lab, history shows that spills/transfers = 4/1036. During the next quarter, 358 transfers are scheduled. Estimate the probability of at least one spill during that time.

3. Flat Tire

Two high-school boys played hooky from school one morning and gave the teacher the excuse that their car had a flat tire. "Well," she said, "each one of you write down on a piece of paper which tire was flat." What is the probability that both boys pick the same tire?

4. Duplicate Birthdays

Suppose you ask n people their birthday—that is, the month and day, but not the year. How large must n be so that the probability of duplicate birthdays is as near as possible to .50? This is the classical birthday problem.

5. Birthday Problem Variation

As a variation on the classical birthday problem, suppose you ask n people just the day of the month when they were born. How large must n be so that the probability of duplicate days is as near as possible to .50?

6. Getting off an Elevator

A hotel elevator starts from the lobby with 7 passengers, and there are 10 floors above with rooms. What is the probability that everyone gets off at a different floor?

7. Your Birthday

Suppose you ask n people their birthday—that is, the month and day, but not the year. How large must n be so that the probability that at least one person has your birthday is as near as possible to .50?

8. General Birthday Problem

This problem was suggested by Norman Thomson, IBM United Kingdom Laboratories Limited, as a "rider" to the third problem last March. In that problem, Problem 7 here, you were asked how many people are required so that the probability that at least one person has your birthday is as near as possible to .50. Norman generalizes the problem as follows. Each member of an indefinitely large population has one of m equally likely values, and we want to find a member with a particular one of the m values. If n is the sample size and $100C\%$ is the desired confidence of at least one success, find an approximate n/m expression that is independent of m. *Hint*: $\ln (1 + x) = x - (x^2/2) + (x^3/3) - \ldots$

9. A Famous French Statistician

The anti-ln of x is e^x. Negate both sides of the n/m equation in the solution to Problem 8, and then take the anti-ln of both sides. What famous French statistician does this remind you of?

10. Finding a Bad Component on a Board

A board has m components. When one of the components fails, the failure analysis people must work their way through the board until they find the bad one. Assume the bad one occurs at random somewhere on the board. If n is the sample size, and $100C\%$ is the confidence that the fail is among the n, find the required n/m.

11. Finding a Bad Circuit in a Component

A complex electronic component has $N = 1000$ circuits. The test equipment engineering people want to use random test patterns of n circuits per component as the components go through the tester. Find n so that the confidence is 99% that a defective component will be detected.

12. Chips on Modules

Chips that are 3% defective are placed on modules, with six chips per module. By definition, a defective module has one or more defective chips. If the defective chips occur at random, what is the percent defective modules?

13. Passing Final Test

At a certain point in the manufacturing line, a unit has one type A defect, three type B defects, and one type C defect. For a single type A defect, the probability of failure at final test is .10. The probability for a type B defect is .20, and it's .80 for a type C defect. If the final test results for the defects are independent of each other, what is the probability the unit passes final test?

14. Leaving a Row

Six students take a statistics test in a row of desks, with an aisle at each end of the row. They leave when they finish the test, in a random order. The instructor notices that all the students are able to go to one of the two aisles without passing by another student still taking the test. He wonders what the probability is of this happening. What is it?

15. A Man and a Woman

This problem was suggested by Leon Lareau, IBM Burlington. A man and woman go separately to the same bar after work for a quick 10-minute drink. If each one arrives at random between five and six o'clock, what is the probability they are in the bar at the same time?

16. A Man and a Woman Waiting

In Problem 15, we saw that if a man and woman go separately to the same bar at random between five and six o'clock, and stay for 10 minutes, then the probability they are in the bar at the same time is only 11/36, or 31%. How long must they stay for a 50% probability?

17. Door Prizes

This problem was suggested by John Walczyk, IBM Boca Raton. For door prizes at a church social, 5 numbered tickets are drawn at random from a total of 64. What is the probability that the numbers are in ascending order?

18. Putting Letters in Envelopes

A secretary has 8 letters ready to go into 8 envelopes, when her boyfriend calls with some upsetting news. After she hangs up, her mind is more on him than the letters, and she ends up putting the letters at random into the envelopes. What is the probability that at least one person gets the right letter?

19. Two Children

At a class reunion, an old pal evasively tells you, "I have two children, and at least one of them is a boy." What is the probability that they are both boys?

20. Two Clocks

At a hotel, you find that the radio part of your digital clock radio doesn't work, although the clock part does. You call hotel maintenance, whose solution is to put in a new clock radio and leave the old one. The maintenance man glances at the time on the old clock and sets the time on the new clock to that. The clock is such that the setting is exact: the desired minutes and no seconds. Some time later you notice that the clock times differ by a minute. What is the probability of this: the clock-setting as described and then disagreement later at a random point in time?

21. Three Vendors

Vendors A, B, and C supply components to Company X.

Vendor	% of Total Quantity	% Defective
A	40	10
B	35	5
C	25	1

What percent of the total defectives are from Vendor A?

22. Changing Partners in a Card Game

This problem was suggested by our editor, Lyle Dockendorf, IBM Rochester. Alice, Bill, Carol, and Donald play 500, a partnership card game. After each game, they switch partners, so that after three games, all possible pairings have been made. What is the probability that somebody (any one of the four) wins all three games?

23. Random Defective Components on Cards

Components that are p fraction defective are placed on cards, with M components per card. By definition, a defective card has one or more defective components. Suppose the defective components occur at random across the cards. For $p = .03$ and $M = 6$, what is the percent defective cards?

24. Clustering of Defective Components on Cards

Now suppose the defective components cluster, to such an extent that the components on a card are either all bad or all good. What is the percent defective cards when $p = .03$ and $M = 6$?

25. One Defective Component per Card

Suppose we have the opposite of clustering, so that a defective card always has just one defective component. For $p = .03$ and $M = 6$, what is the percent defective cards?

26. Shrimp Chow Mein

At the company cafeteria, you get just one shrimp in your shrimp chow mein. You wonder if you are the victim of chance or corporate cost-cutting. Suppose the average number of shrimp per serving is a satisfactory 5. What is the probability that you get one or none?

27. Outgoing Percent Defective After 100% Screen

Problems 27–29 were all suggested by Curtis Jones, IBM San Jose, after he read "Bayes' Theorem and Screening for Future Mental Illness: An Exercise in Preventive Community Psychiatry" by Lyon Hyams, as reproduced in Scherr (14).

A certain reliability defect can be detected only with a destructive test. However, Quality Engineering has developed a nondestructive electrical test such that product units with the reliability defect fail the test 95% of the time. Further, units without the reliability defect pass the electrical test 95% of the time. Quality Engineering proposes using the electrical test as a 100% screen on all the product. Units that fail the screen will not be replaced. If 2% of the product has the reliability defect, what will be the outgoing percent defective for the reliability defect after the screen?

28. Yield for 100% Screen

Manufacturing Engineering wants to know what the yield will be for the screen when 2% of the product has the reliability defect. In other words, what percent of the units pass the screen?

29. Quality of Scrap for 100% Screen

If 2% of the product has the reliability defect, and we collect all the defectives from the screen, what percent of them would actually have the reliability defect?

30. Even Number of Occurrences

Consider the binomial distribution model with n trials, p probability of occurrence on each trial, $q = 1 - p$, and x occurrences in the n trials. Find the probability that x is even, without having to sum up all the probability terms for the even x values. Treat zero as an even number.

31. Bit Lines and Vias

A chip has 1152 bit lines, and each bit line has 256 vias. By definition, a bad bit line has 2 or more bad vias, and a bad chip has one or more bad bit lines. A reliability test gives only the number of bad bit lines on a bad chip. For a particular sample of 331 chips, there are 3 bad chips, and each bad chip has exactly one bad bit line. Estimate the overall fraction bad vias.

COMBINATIONS AND PERMUTATIONS

32. Clusters

Cluster analysis is a technique for classifying objects into clusters, such that objects in the same cluster are similar to one another. With the divisive method of cluster analysis, the first step is to divide the initial group of objects into two. How many ways can this be done for 10 objects?

33. NCAA Basketball Tournament

For the 1983 NCAA basketball tournament, the 52 qualifying teams were reduced to one through a system of games and byes based on seeding. Only the 8 lowest-ranked teams had to play in the opening round. The 4 winners there advanced to the first round along with 28 others who received byes to that round. Then the 16 first-round winners went to the second round with the 16 top-ranked teams, who got byes all the way to the second round. There were no more byes after that. How many games were played in the whole tournament before North Carolina State was finally crowned as the national champion?

34. Password

An APL password can be four alphanumeric characters, with duplicate letters or digits allowed. There must be a mixture of letters and digits, so ABCD and 1234 are not valid. How many such passwords are there?

35. Trimming a Tree

This problem is from Leon Lareau, IBM Burlington. The path shown on the tree connects 12345678 in order. With only right-angle turns allowed, how many such paths are there?

```
        8
       878
      87678
     8765678
    876545618
   87654345678
  876543234 3678
 8765432 1 2 345678
```

36. Combinations of Chemicals

Problems 36–38 were suggested by Lyle Dockendorf, IBM Rochester. A chemist wishes to test the effect of adding any combination of four chemicals to a given solution. How many different combinations are there, including the case of no additions?

37. Combinations of Chemicals (*continued*)

In general, a combination has r chemicals present, where $r=0$, 1,2,3,4. For what value(s) of r is the number of possible combinations a maximum?

38. Testing All Combinations of Chemicals

In order to test all the combinations of chemicals, the chemist could take a separate sample from the given solution for each combination, add the combination, and test. However, this isn't really necessary because he can add and test as he goes along. For example, with just one sample, he could add chemical A, test, add chemical B, test, etc., to cover combinations A, AB, etc. What is the smallest number of samples that could be taken and still allow all the combinations to be tested?

39. Seating Diplomats

Six diplomats must be seated at unmarked places around a circular conference table. How many possible arrangements are there?

40. Two Packs of Life Savers

You have two packs of Live Savers: Butter Rum and Wild Cherry. Each pack has 11 candies. If you take them one at a time from the ends of the packs, how many different orders are there for comsuming them?

41. Life Savers Puzzle

Life Savers has come out with a puzzle composed of 12 interlocking plastic pieces (oversized Life Savers) that will fit together only one way. Each piece has two different sides. If, just using trial and error, you stack the pieces in all possible orders, how many stacks are there to try?

42. Driving to Work

Problems 42 and 43 are from Lyle Dockendorf and Arie Lagerwaard, both from IBM Rochester. You live in a city with a grid of streets and avenues. All blocks are squares of the same size. Each morning you drive from your apartment at the intersection of 2nd St. and 2nd Ave. to your office at the intersection of 10th

St. and 10th Ave. Your route has always been eight blocks to the north, and then eight blocks to the east. Then one day you realize that you could put a little variety in your life by taking different routes. If you don't drive any farther than necessary, how many equal-distance routes can you take to work?

43. Walking to Work

In Problem 42, you would like to take a closer look at things along the way, so you decide to *walk* rather than drive. Your apartment is at the southwest corner of the 2nd St. and 2nd Ave. intersection, and your office is at the southwest corner of the 10th St. and 10th Ave. intersection. Different sides of a street or avenue make the route different, as do different ways to cross an intersection, but no diagonal crossing is allowed. If you don't walk any farther than necessary, now how many equal-distance routes can you take to work?

44. Matching Socks

You have 7 pairs of socks: 5 pairs of one kind and 2 of another. One day after a wash, you don't feel like matching the socks, so you just throw them in the drawer. The next morning you reach in the drawer without looking and take out 2 socks. What is the probability that you have a matching pair?

45. Matching Socks (*continued*)

In Problem 44 suppose you draw two socks from the drawer without looking each day for the whole week. What is the probability of getting a match on every one of the 7 days?

46. Choosing Desserts

Suppose there are *n* different items on a dessert tray, and you may take as many as you like. How many total sets of 0, 1, 2, . . . desserts must you consider?

47. 5-Button Combination Lock

A door has a 5-button combination lock on the knob. The buttons are numbered 1 through 5. For a combination, suppose there can be 3 or 4 entries, and one of these has to be a double like 13, which is 1 and 3 together. Once a digit is used, it can't be used again. As examples, a three-entry combination might be 14, 3, 2, and a four-entry combination might be 2, 15, 4, 3. If you don't know the combination, what is the probability you guess it right on the first try?

48. Fish on Venus

On Venus there is a fish type that has three sexes, and a mated triplet is necessary to produce offspring. To an astronaut, the fish are indistinguishable by sex. How many fish must an astronaut bring back so that the probability of having at least one fish of each sex is over .5?

GAMES

49. Blackjack

In the casino game of blackjack, or 21, players initially receive two cards from four well-shuffled decks. If the cards are an ace and a 10 (10, jack, queen, or king), this is 21, and the player receives 1.5 times his bet. Suppose a 21 occurs once every n hands on the average, and find n to the nearest integer.

50. Winning at Craps

In the casino game of craps, the player rolls two dice. If the first roll is 7 or 11, he wins, but if it's 2, 3, or 12, then it's "craps," and he loses. Any other total on the first roll becomes the "point," and the player continues to roll the dice until he either gets the point, in which case he wins, or gets a 7, in which case he loses. Totals other than the point or 7 don't count. Note that 7 is good on the first roll, but bad on later rolls. What is the probability that the player wins?

51. 6 Before 7

In the casino game of craps, if the shooter's first roll is a 6, then he must keep rolling until he gets a 6 and wins, or a 7 and loses. Totals other than 6 or 7 don't count. What is the probability of getting a 6 before a 7?

52. Like Roulette?

From a deck of playing cards, take 18 reds, 18 blacks, and the 2 jokers. You could shuffle these cards and pretend you're playing roulette by dealing them one at a time: red you win, black or joker you lose. The jokers correspond to the house zero and double-zero. Suppose you think of the house as having $100, and as you go through the deck you always bet 20% of the house's money. This means you bet $20 on the first card, then .20 (100 + 20) or .20 (100 − 20) on the second card, etc. What can you expect to win on one run through the deck?

53. Winning a Kewpie Doll

At the county fair, there is a game of chance in which a wheel with 20 numbers is spun, and if the wheel stops at your number, you win. Each bet is $.25, and when you win, you get a Kewpie doll. On the average, how much do you pay to get a doll?

54. Pregame Coin Flips

Sports Illustrated reports that a high-school football team in Bloomington, Indiana, lost 21 straight pregame coin flips before finally winning one! We all know the probability of this for a *particular* team is $(1/2)^{21}$, but there are many, many teams. In fact, according to a reliable source, there are approximately 15,000 football teams in the United States, when we consider high school, college, and pro teams. What is the probability today that at least one team has lost the last 21 flips?

55. Beer Tasting

You fancy yourself a connoisseur of beers, but your friend doubts if you can tell the difference among brands. He challenges you to a game in which you try to identify the beers in four glasses. He even tells you the four brands. All you have to do is tell which is which. For every one right he pays you a dollar, and for every one wrong you pay him a dollar, but you don't know the results until you have made all your guesses. You reason that even if your taste buds aren't as great as you think they are, it still should be a fair game, like flipping coins, since you are either right or wrong on each glass. You play the game. If you actually have no ability at all to distinguish among beers, what is the probability you are ahead by some amount of money at the end of the game?

56. Beer Tasting (*continued*)

In Problem 55, a beer-tasting game was described in which you try to identify four different beers in glasses. You are given the brand names. For every one right you win a dollar, and for every one wrong you lose a dollar, but you don't know the results until you have made all your guesses. We saw that if you just guess without using your taste buds, your chance of being ahead by some amount of money at the end of the game is only 1/24. This in itself does not make the game unfair to you. The game is unfair only if your average winnings per game is negative. What is it?

57. Three Cards

This problem is from Lyle Dockendorf, IBM Rochester. A man shows you 3 cards. The first has an X on each side, the second a Y on each side, and the third an X on one side and a Y on the other. He mixes these cards and places one on the table. The side up shows a Y. He suggests a bet: "I'll wager $1.50 to your $1.00 that the other side of that card is also a Y." Do you take the bet?

58. Choke?

Let's go back to game four of the 1984 National Basketball Association championship series between the Celtics and the Lakers. With the score tied and only 35 seconds left in overtime, Earvin

"Where Has the Magic Gone" Johnson steps to the foul line for two shots. He misses them both! Statistics show that Earv is an 80% foul shooter. What is the probability that he missed those two shots just due to chance?

59. Another Choke?

This one is for all you Laker fans. In the first game of the 1985 National Basketball Association championships series between the Celtics and the Lakers, Celtic guard Danny Ainge shot 9/15 = 60% from the floor, but in the sixth and final game he was 3/16 = 19%. Can this difference be reasonably attributed just to chance?

60. Repeated Number in Lottery

The winning three-digit number in a state lottery is determined each day by a random selection of three ping-pong balls, with the digits 0 and 1 through 9 equally likely to occur on a ball. One day the number is the same as the one on the previous day! The newspapers make a big thing about it, and the state has the selection procedure checked, but everything is okay. The lottery has been running for five years, with a number every day except Sunday. What is the probability that during that long a period there will be at least one time when today's number is the same as yesterday's?

61. House Edge for Lottery

For a state lottery, a three-digit number is picked each day, with all numbers from 000–999 equally likely. If you buy a $1 ticket and win, the state pays you $500. Find the "house edge," which is defined as the expected long-run gain by the state, expressed as a percent of the amount bet.

62. A Statistician's Urn

A statistician's urn contains 31 white balls and 469 black ones. He takes out two balls at random and replaces them with one black if they match in color, or one white if they don't. He keeps doing this until only one ball remains. What color is it?

63. Another Statistician's Urn

This problem is from Charlie Magill, IBM Burlington, who got it from *The Bent of Tau Beta Pi* magazine, who got it from Hermann Laurent, 1873. An urn initially contains one white ball and one black ball. A trial is defined as two draws by a player, with replacement after each draw. If both draws are white, the player wins, and the game is over. Otherwise, a black ball is added to the urn, and the process is repeated. This continues indefinitely. Find the probability that the player wins.

64. Yet Another Statistician's Urn

Allan Atrubin, IBM Rochester, offers the following variation on Problem 63. An urn initially contains one white ball and one black ball. A trial is defined as one draw by a player, with replacement after each draw. If the draw is white, the player wins, and the game is over. Otherwise, a black ball is added to the urn, and the process is repeated. This continues indefinitely. Find the probability that the player wins.

65. Cutting for the Deal

To start the two-person card game of cribbage, one player cuts the deck to get a card, and then the other player cuts what is left to get his card. The player with the lower card deals. Ace is the lowest possible card, and then 2, 3, 4, etc. One day, two players both cut an ace! They wonder what the probability is of this happening. What is it?

66. Hexogram

The game of Hexogram is played with triangular pieces, and each piece is divided into three colored sections as shown. There are

5 colors in all, and no two pieces are the same. How many possible pieces are there?

67. Yarborough in Bridge

According to legend, the Earl of Yarborough offered a bet of 1000 to 1 that a bridge hand would have no card higher than a 9. That's why such a hand is called a Yarborough. In bridge, ace is high, so the 13 cards in a Yarborough are all 2 through 9. What are the fair odds against a Yarborough?

68. Perfect Cribbage Hand

In the card game of cribbage, you are dealt six cards. You keep four of these and the other two go into the "kitty." Then the deck is cut for the fifth card. It turns out that the best possible hand is 29, which is three 5s and a jack in your hand, with the cut-card equal to 5 and the same suit as your jack. What is the probability of a perfect hand?

69. Probability of Player's Ruin

Suppose a craps player starts with $900 and plays until he either increases it to $1000 or has nothing left. If the bet size is $1, what is the probability of the player's "ruin?"

70. Probability of Player's Ruin (*continued*)

Suppose the player in Problem 69 increases the bet size to $100. Now what is the probability of the player's ruin?

71. Celtics Home and Away Records

For the '86–'87 season, the Boston Celtics were 39–2 at home and 20–21 away. Except for two odd games, the season can be thought of as 40 pairs of home and away games, with a common opponent for each pair. Results for these 40 pairs were as shown.

	Home	
Away	Win	Lose
Win	18	1
Lose	20	1

What is the probability of such a big difference in the home and away performance, in the observed direction, just due to chance?

72. Celtics Home and Away Stats

Apparently, the Celtics *do* perform better at home than away. Why? For the 40 pairs in Problem 71, the Celtics statistics were as shown. Fill in the one-tail probability column to find which statistics show a significant difference at the .05 level.

Statistic	Mean			Sign test			
	Away	Home	Home–away	Home–away −	0	+	One-tail probability
Field-goal %	50.0	53.6	3.6	12	0	28	
Free-throw %	79.1	82.1	3.0	16	0	24	
Rebounds	48.0	51.6	3.6	13	1	26	
Assists	28.0	31.0	3.0	12	2	26	
Personal fouls	22.0	20.1	−1.9	26	2	12	
Technical fouls	.8	.6	−.2	11	20	9	
Total fouls	22.8	20.7	−2.1	26	1	13	

73. Two-Time Loser

Six people play a coffee game in the morning, with the loser buying for everybody. That afternoon they play again. What is the probability that the same person loses both times?

74. Russian Roulette

In the solitaire game of Russian roulette, the player puts a bullet in the six-chamber cylinder of a pistol, spins the cylinder, points the pistol at his head, and pulls the trigger. Suppose a player decides to use two bullets and two trigger pulls. Now he has choices. He can put the bullets in consecutively or spin the cylinder between placements. Then he can use consecutive trigger pulls or spin the cylinder between pulls. Find the survival probabilities for all the different ways he can play the game.

75. Momentum

In the 1987 World Series, the St. Louis Cardinals won Game 5 to give them a three-games-to-two edge over the Minnesota Twins. According to the Associated Press, "The numbers of history support the Cardinals and the momentum they carry. Whenever a World Series has been tied 2–2, the team that won Game 5 was eventually the champion 71 percent of the time." If "momentum" is not a factor, and each team has a 50% chance of winning a game, what is the probability that the Game 5 winner eventually wins the Series?

76. Sudden Death

In the National Football League, ties are settled with a sudden-death playoff. The receiving team is decided with a coin toss, and then the first team to score wins the game. Suppose the probability of scoring on a possession is .5 for each team. What is the probability that the receiving team wins the game?

77. Rolling a Die

Suppose you are rolling a six-sided die until you get all six numbers at least once. You already have four of the numbers: 1, 2, 4, and 5. On the average, how many rolls do you make before you get the next new number?

78. Rolling a Die (*continued*)

In Problem 77, if you start from scratch, what is the average number of rolls to get all six numbers at least once?

79. Six Different Numbers in Six Rolls

In Problem 78, what is the probability that you get all six numbers after just six rolls?

80. A Rich Uncle

Every time a rich uncle visits you, he brings two envelopes, one containing twice as much money as the other. He lets you pick an envelope and look at the contents. Then you can either keep the envelope or trade for the other. Unfortunately, he varies the amounts in the envelopes from visit to visit, so knowing the contents of one envelope doesn't tell you what's in the other one. You want to make as much money as possible in the long run. Should you trade?

81. A Rich Aunt

You also have a rich aunt, and every time she visits you, she gives you an envelope with some money in it. Then she gives you a chance to trade for another envelope that has either half as much money or twice as much money, with the same .5 probability for each of these possibilities. Again, you want to make as much money as possible in the long run. Should you trade?

82. Progression in Football Standings

In January 1989, someone wrote in to *Sports Illustrated* to note the "perfect mathematical progression" in the Big Eight's final standings for the football season: 7–0, 6–1, 5–2, 4–3, 3–4, 2–5, 1–6, and 0–7. The reader wanted to know the probability of such a progression happening. *SI* said, "If you could assume that all teams had a 50–50 chance of winning each game they played and that every game produced a winner, the probability of this progression occurring in an eight-team conference would be 1 in 6,667." Can you produce the 6,667?

83. Bingo Cover-All

The game of Bingo is played with cards, each of which has a 5-by-5 square of 25 numbers, except the middle space, which is Free. From left to right, the columns are BINGO. On a card, the columns have a random deal of numbers, with no duplicates, as follows: 5 from 1–15 in B, 5 from 16–30 in I, 4 from 31–45 and the Free space in N, 5 from 46–60 in G, and 5 from 61–75 in O. A player covers up the Free space at the beginning of the game. Then the caller draws balls one at a time from a drum of well-mixed balls numbered from 1 through 75. If a player has the number called, he covers it up. Usually, the object is to get Bingo: a string of 5 numbers covered in a horizontal, vertical, or diagonal direction. However, once a night at a particular club, they have a Cover-All, with the object being to cover the whole card by the time 55 numbers have been called. What is the probability that a particular card is a winner at one Cover-All game?

84. Bingo Cover-All (*continued*)

In Problem 83, we saw that the probability of a Bingo Cover-All (all 24 numbers covered) in 55 numbers called is .00010. Suppose the caller continues after the 55th number until all 75 numbers have been called. Then you are sure to get a Cover-All at some point. What is the probability that you have to wait until the 75th number to get it?

SAMPLING

85. Random Sample

At wafer foreign-material inspection, a boat contains up to 25 wafers. The inspector knows that wafers toward the ends of the boat have more particles, so his "random" selection of wafers for inspection is even more difficult than it ordinarily would be. Find a simple method to take a random sample of n wafers out of N, where $N \leq 25$.

86. Sampling Tool

To show his statistics class that sample results vary, a teacher uses as a sampling tool a small drum with $N = 1000$ balls, 1% of which are yellow, 2% red, etc. Turning the drum one revolution mixes the balls, and a random sample appears in the window. For sample size $n = 80$ and $x =$ number of yellow balls, results for two days were as shown.

	Number of samples	
x	Day 1	Day 2
0	17	15
1	24	25
2	6	9
3	2	1
4	1	0

The teacher is puzzled by these results because he expects a binomial distribution with $n = 80$ and $p = .01$, which has the peak probability at $x = 0$, not $x = 1$. Is something wrong with the sampling tool?

87. Sampling Tool (*continued*)

The teacher in Problem 86 looks through the window of the sampling tool to see the balls as the drum turns. He notes that the balls don't really mix together very much. Balls on top tend to stay on top. In effect, then, the sample in the window is from some population size $N < 1000$ with the population of k yellow balls. Find an N and k consistent with the Problem 86 data.

88. Marked Fish

Suppose 100 marked fish are released in a lake. Later we catch 50 fish, and 5 of these are marked. What is the estimate for the total number of fish in the lake just before the 50 were caught?

89. Sampling as You Go Along

In the July 1987 *Satishare*, Lyle Dockendorf, IBM Rochester, presented a clever sampling procedure for sequentially selecting a random sample of fixed size from a population whose size will not be known until the end of the sampling period. For example, you may want to fill a stress chamber with units for a weekly reliability test, but the population size for the week is unknown until the end of the week. Lyle shows how to randomly sample as you go along, so that all units have the same probability of being in the sample. With Lyle's permission, I am simplifying the notation a bit. Let n be the sample size, and N be the unknown population size. The first n units are tentatively taken as the sample. For the ith unit, $i > n$, select a random number u from the uniform distribution between 0 and 1. Let j be the next integer greater than or equal to iu. If $j \leq n$, the ith unit replaces the prior jth unit in the sample. Otherwise, the ith unit is not in the sample. Prove that the first n units have the same probability of being in the sample.

90. Sampling as You Go Along (*continued*)

In Problem 89, show that the units *after* the nth unit all have the same probability of being in the sample as the first n units.

91. Side-by-Side

Product units are stored in tubes, with 20 units/tube and 100 tubes/lot. Since random sampling for lot acceptance would require unloading many tubes and then reloading them, engineers like to claim that the defectives are spread at random throughout the lot. This justifies taking a small number of tubes and inspecting all the units in those tubes. A skeptical statistician convinces management to have the inspection results recorded sequentially for 10 tubes from a particular lot. With positions 1–20 for the first tube, 21–40 for the second, etc., defectives were found only in positions 83 and 84. If two defectives are placed at random along the 200 positions, what is the probability that they end up side-by-side like this?

92. Randomness Test for 0s and 1s

In the previous problem, suppose 10 tubes from another lot are inspected, and defectives occur in positions 46, 81, 82, 111, 139, 140, 141, 164, 182, and 183. Consider the string of 200 0s and 1s, with 0 for nondefective and 1 for defective, and define a run to be a sequence of all 0s or all 1s. The total number of runs can be used to check for randomness, since if there is clustering of defectives, the number of runs will be small. Find the probability here of the observed number of runs or fewer.

93. Runs Distribution

See the solution to the previous problem. Many texts say that when m and n are both 10 or more, the distribution of u is approximately normal, and the exact probability formulas need not be used. When $m = 10$ and $n = 190$, as here, find the u probability distribution with the exact formulas. Does it look normal to you?

94. Randomness Test for Normal Distribution

For a certain electrical test, defective parts per million (ppm) for the last six months have been: 1171, 1846, 514, 1618, 400, and 1000. Test the hypothesis that this data shows random variation from a normal distribution. Are there too many ups and downs?

95. Components of Variance

For a certain dimension x, a random sample of $m = 4$ units from each of $n = 20$ large lots has been measured. Estimate the percent of the total process variance that is due to the variance of the lot means, if you are given the following information. The variance of the *sample* means, with $\bar{x} = \Sigma x/m$ and $\bar{\bar{x}} = \Sigma \bar{x}/n$, is:

$$s_1^2 = \frac{\Sigma(\bar{x} - \bar{\bar{x}})^2}{n - 1} = 6.688$$

The within-sample variance for the ith lot is calculated from:

$$s_{2i}^2 = \frac{\Sigma(x_i - \bar{x}_i)^2}{m - 1}$$

The mean within-sample variance for the $n = 20$ lots is:

$$s_2^2 = \frac{\Sigma s_{2i}^2}{n} = 2.933$$

96. Components of Variance (*continued*)

In the previous problem, the engineers wonder what the three-sigma for x would be if process changes make future lot means virtually the same. What do you say?

97. Balls in Cells

Within a certain manufacturing tool, 20 heads make product units, which all fall into the same bin. Periodically, an inspector checks a random sample of 32 units from an indefinitely large number of units in the bin. Find the mean (expected) number of heads with no units in the sample.

SURVEYS

98. Poll Before Election

Before a city election for mayor, a radio station takes a poll of 100 voters at random. The results are 30% for Adams, 20% for Brown, 40% for Carr, and 10% undecided. Find the probability of such a large difference between Adams and Carr, just due to chance.

99. Opinion Survey

For the IBM Opinion Survey, percents favorable on a certain question are 71, 64, 68, 79, and 69 for organizations of 119, 430, 169, 266, and 433 people. Management wants to know if these differences can be reasonably attributed just to chance. Are the differences significant at the .05 level?

100. Error for Survey

Personnel plans to run an opinion survey to see how the employees in a department feel about their manager. Questions will be of the "yes" or "no" variety. Suppose the population size is N and the sample size is n. For 95% confidence, find an expression for the maximum difference between the sample and population proportions "yes" for a certain question. This expression should be a function of n and N only. Hint: Use the normal distribution approximation to the hypergeometric distribution. Something nice happens if we approximate the exact $z = 1.96$ with 2.

101. Sample Size for Survey

A company would like to run a sample survey of its stockholders to see how they feel about certain issues. Questions will be of the "yes" or "no" variety. If there are $N = 10,000$ stockholders, how large must the sample size n be in order to have 95% confidence that the sampling error is less than $\pm 5\%$? Use the normal distribution approximation to the hypergeometric distribution, and replace the exact $z = 1.96$ with 2.

102. Nonresponse in Survey

A company runs a survey of its stockholders to see how they feel about certain issues. Questions are of the "yes" or "no" variety. Unfortunately, only $n = 200$ of the $N = 1000$ total stockholders respond to the questionnaire. For a particular question, 90% of the respondents answer "yes." With 100% confidence, find lower and upper limits for the percent answering "yes" in the whole population of $N = 1000$.

103. Nonresponse in Survey (*continued*)

In Problem 102, suppose the 1000 people who were sent the questionnaire are a random sample from an indefinitely large population. Now find 95% confidence limits for the percent "yes" in the whole population. Ignore the possible difference between the $200/1000 = .2$ *sample* proportion responding and the *population* proportion responding if everyone were sent a questionnaire.

POPULATION PERCENT DEFECTIVE

104. Percent Out-of-Spec

A certain dimension follows a normal distribution with a mean of $\mu_x = 161.2$ and a standard deviation of $\sigma_x = 5.6$. The lower specification limit for individual readings is LSL $= 150$, and the upper specification limit is USL $= 170$. Estimate the percent defective and sketch the distribution.

105. Moving the Mean

For the process in the previous problem, the nominal value for the dimension is midway between the LSL and USL, or 160. If adjustments were made to the process so that $\mu_x = 160$, what would be the new percent defective?

106. Adding and Subtracting Variables

A critical dimension y for a unit is a function of three other independent dimensions:

$$y = x_1 + x_2 - x_3$$

35

For the x_i variables, means and standard deviations are as shown.

i	μ_i	σ_i
1	39.06	.42
2	5.64	.23
3	12.32	.16

Find the mean and standard deviation for y.

107. Defect Density to Achieve Percent Defective

Suppose defects occur at random across a product unit, and the total area of the unit is 2 square inches. Find the defect density expressed as defects per square inch such that the percent defective units is only 1%.

STATISTICAL PROCESS CONTROL (SPC)

108. Percent Defective Stable?

For a certain visual inspection, a random sample of 200 units per lot is inspected. Defective units by lot for the last 20 lots are: 3, 2, 2, 1, 4, 2, 3, 2, 2, 0, 2, 1, 5, 1, 3, 2, 2, 2, 4, and 2. Do you think these differences can be reasonably attributed just to chance sampling fluctuations?

109. Defects per Unit Area Stable?

This problem was suggested by Gary Snyder, IBM Burlington, for random defects on a product unit. Suppose c is the variable number of defects found in an area A, but only A and the defects per unit area c/A are input to a computer system. If $A = 40$ square centimeters and the mean value of c/A is .52, find three-sigma control limits for c/A.

110. Standard Deviation Stable?

For statistical process control of a certain dimension x, Quality Control periodically measures a sample of $n = 4$ product units. The sample range R is the highest reading minus the lowest. For a history of 20 lots, R has fluctuated at random between 1.3 and 6.5, with an average of $\overline{R} = 3.665$. Is the process standard deviation σ_x stable?

111. Mean Stable?

In Problem 110, the sample average $\overline{x}$ fluctuates at random between 5.82 and 15.42, with an average of $\overline{\overline{x}} = 10.79$. Is the process mean μ_x stable?

112. Process Meeting Spec?

The control limits in Problems 110 and 111 are the classical ones, designed to answer the question, "Is the process stable?" The specification for x is not taken into account. Suppose the upper specification limit (USL) for x is 20 and the lower specification limit (LSL) is 2. Let's agree that the process is meeting the spec if the three-sigma limits for the x distribution are within the spec. Now answer the question, "Is the process meeting the specification?"

113. p-Chart ARL Equals Cusum ARL

Two methods of process control for a percent defective are p-Chart and Cusum. For a p-Chart, we take a sample of n units and accept if the number of defectives is c or less. Under a Cusum plan, the sample defectives x are used to calculate a cumulative sum or cusum: $S = \Sigma(x - k)$. We reject $S > h$, accept on $S \leq h$, and change a negative S to zero. The goodness of a process control plan can be assessed by its average run length (ARL) curve, which gives the average number of points to rejection as a function of process quality. In order to compare plans, we can overlay the ARL curves with GRAFSTAT, an APL interactive data analysis system. When we do this for p-Chart and Cusum, we occasionally may be surprised to see that the curves coincide. Under what conditions does this happen?

114. p-Chart ARL Greater than Cusum ARL

Under what conditions does the p-Chart ARL curve lie *above* the Cusum ARL curve?

115. p-Chart ARL Less than Cusum ARL

Under what conditions does the p-Chart ARL curve lie *below* the Cusum ARL curve?

116. ARL for Control Limits

Suppose you are using an $\bar{x}$ chart to control the process average for a product variable x. You have a solid central line for the target process average, and dashed lines for the three-sigma control limits around the target. Chapter 21 of Duncan (5) shows how to set up the chart. A "signal" is a sample $\bar{x}$ outside the control limits. If the process average is stable at the target, find the ARL = average run length, the average number of $\bar{x}$ values to get a signal.

117. ARL for Rule of Seven

According to the Rule of Seven [Duncan (5), p. 432] for control chart analysis, a point that completes a run of seven consecutive points on one side of the target is also a signal, even if all the points are within the control limits. In Problem 116, find the ARL for the Rule of Seven by itself when the process average is stable at the target. Ignore the control limits.

118. ARL for Control Limits and Rule of Seven

In Problem 116, suppose you use the control limits *and* the Rule of Seven from Problem 117. Now you have two ways to get a signal: outside the control limits or a run of seven. What is the ARL when the process average is stable at the target?

MEASUREMENT ACCURACY

119. Measurements on a Standard

A standards lab measures a unit 8 times on its tester, and the readings in order are: 25.2, 25.8, 26.1, 26.2, 25.9, 25.8, 25.7, and 25.0. The "standard value" is an estimate of the true value. Find the standard value and its three-sigma error.

120. Standard Value and Error

After maintenance work on the standards lab tester in Problem 119, a repeatability study shows that readings on the same unit now fluctuate at random over time. For a particular standard, the readings in order are: 26.1, 26.1, 26.0, 25.9, 26.2, 26.0, 25.9, and 26.0. The "standard value" is an estimate of the true value. Find the standard value and its three-sigma error.

121. Tester Bias

As a check on a line tester, $n = 20$ readings are taken on a standard over a two-week period. The readings fluctuate at random about the sample mean of $\bar{x} = 26.18$, with three sample standard deviations of $3s_x = 1.43$. The standard value with its three-sigma error is $26.03 \pm .11$. Find the bias estimate and its three-sigma error, where bias is defined as the line tester's long-run mean minus the standards lab's.

QUALITY CONTROL AUDITS

122. Missing Defects

For a difficult widget inspection, the Manufacturing inspector is allowed to miss 10% of the defects in the long run. Nobody is prefect. During an audit over a period of time, the Quality Control inspector looks at the same widgets, good and bad, and finds 25 defects. Suppose we want the Manufacturing inspector to have at least a 95% chance of acceptance if he is really at the 10% acceptable value in the long run. How many of the 25 defects do we allow him to miss?

123. Measurement of Variable on Same Units

A Manufacturing inspector routinely uses Method A to measure a critical dimension on a product unit. For an audit, a Quality Control inspector periodically checks the measurement with Method B. Measurement accuracy studies show that Method A has a three-sigma repeatability error of $\pm .372$, while the value for Method

B is $\pm .100$. Find the three-sigma control limits for the difference between the two measurements.

124. Measurement of Variable on Independent Sample of Units

A Manufacturing inspector takes a random sample of $n_1 = 5$ product units from a lot, measures them for a dimension x, and finds the mean $\bar{x}_1 = 4.84$ and the standard deviation $s_1 = .15$. For an audit, a Quality Control inspector takes an independent random sample of $n_2 = 5$ units, measures them, and finds $\bar{x}_2 = 4.16$ and $s_2 = .79$. Find the probability of this great a difference or more in the sample means just due to chance.

125. Percent Defective for Same Units

At a difficult visual inspection, a Manufacturing inspector looks at a random sample of 50 units from a lot and separates them into goods and bads. For an audit, a Quality Control inspector looks at the same units to check the correctness of the inspection. The QC inspector is the check inspector, so if he says it's good, it's good. A history of audits shows that the following inspection errors are the best that can be done in the long run:

$\alpha = $ Fraction of true goods called bad by Manufacturing $ = .01$

$\beta = $ Fraction of true bads called good by Manufacturing $ = .10$

Suppose for each audit of 50 units, we want to run significance tests on the misses such that the Manufacturing inspector has at least a 95% chance of acceptance if he is doing $\alpha = .01$ work on the true goods, and also at least a 95% chance of acceptance if he is doing $\beta = .10$ work on the true bads. Prepare a table of allowable misses that could cover the different cases as the QC goods goes from 45 to 50.

126. Percent Defective for Independent Sample of Units

A Manufacturing inspector takes a random sample of $n_1 = 125$ units from a lot, inspects them visually for defects, and finds x_1

= 0 defective units. For an audit, a Quality Control inspector takes an independent random sample of n_2 = 125 units and finds x_2 defective units. How large must x_2 be in order to say there is a significant difference between the two inspectors? Use a one-tail test at the α = .05 significance level.

127. Sampling After Imperfect 100% Inspection

After 100% inspection by Manufacturing, a small lot has 10 supposedly good units left, but 2 of them are actually defective, due to inspection error. For an audit, the Quality Control inspector takes a random 5 units out of the 10. What is the probability he takes all nondefectives?

SAMPLING PLANS

128. Acceptable Quality Level

A critical product variable is very time-consuming to measure, so a sample of only $n = 6$ units is spread across a week's production. Quality Engineering wants to reject the week if any defectives are found, which means the acceptance number is $c = 0$. Find the acceptable quality level (AQL), which is the population percent defective such that the chance of acceptance is 95%.

129. Acceptable Quality Level vs. Lower Limit

Manufacturing Engineering doesn't think it's right to reject a week for just one defective. They understand the AQL concept, but they also want to know the 95% confidence one-sided lower limit for the week's percent defective, when just $x = 1$ defective is observed in the $n = 6$ units. What is it?

130. Acceptable Quality Level vs. Significance Test

Manufacturing Engineering is still not happy. They want to run a significance test every week to see if the x defectives in the $n = 6$ units is significantly worse than the AQL of .85%. Find the value of x that is just significant at the $\alpha = .05$ level of significance.

131. Good and Bad Quality

For a lot of acceptance sampling plan, a random sample of n units is inspected, and the lot is accepted if the number of defectives is less than or equal to the acceptance number c. Find n and c so that the plan accepts 95% of the lots when the process is 2% defective, but only 10% of the lots when the process is 8% defective.

132. Widgets in Tubes

A particular lot of widgets consists of 561 tubes, with 50 widgets per tube. Due to a gross processing error, 58 of the tubes would be 100% defective for the reliability test, while the other tubes are defect-free. This is not known by the Quality people. The sampling plan accepts on zero defective in a sample of 400 widgets, which is selected by taking all the widgets in a random sample of 8 tubes. What is the probability of acceptance?

133. Zero Defects in Sample

A random sample of 50 units from a lot of 1000 units has zero defects. Find the upper limit for the lot percent defective units, with 90% confidence.

134. LTPD and Upper Limit

A lot acceptance sampling plan calls for a random sample of 80 units, and acceptance on 2 or fewer defectives. The lot tolerance percent defective (LTPD) is determined to be 6.5%, which is the lot percent defective such that the chance of acceptance is only 10%. Find the upper limit for the percent defective units in an accepted lot, with 90% confidence.

135. Average Outgoing Quality for Sampling Plan

For a certain visual inspection, the lot size is $N = 400$ units, and the sample size is $n = 50$ units. A lot is accepted if there are no defectives in the sample, which means the acceptance number is $c = 0$. Rejected lots are screened 100% for defectives. Defectives found during sampling or screening are replaced with nondefectives. Suppose the percent defective incoming to the sampling plan is stable at .20%. Find the average outgoing quality (AOQ), which is defined to be the long-run percent defective for the mix of accepted and screened lots that leave the inspection station.

136. Average Outgoing Quality Limit for Sampling Plan

The average outgoing quality limit (AOQL) is the maximum AOQ over all possible incoming percents defective. Dodge and Romig (4) use a Poisson approximation to the binomial distribution to show that

$$\text{AOQL} = \frac{100y}{n}\left(1 - \frac{n}{N}\right),$$

where y is a tabulated factor, found also in Duncan (5), p. 376. Pretend you don't know about this and find the AOQL for the Problem 135 sampling plan without using the Poisson approximation.

137. Average Outgoing Quality for Imperfect 100% Inspection

A process is so poor that management decides to do 100% inspection, and in the long run 5% of the units are called defective and removed. Independent quality control audits for this inspections show that the errors are:

α = Fraction of true goods called bad = .01

β = Fraction of true bads called good = .10

Estimate the average outgoing quality (AOQ), which is the *true* percent defective after screening.

138. AQL and AOQL Requirements

For a lot acceptance sampling plan, a random sample of n units from a total of $N = 2000$ units is inspected, and the lot is accepted if the number of defectives is less than or equal to the acceptance number c. Find the n and c that most nearly satisfy the following requirements:

   a.   95% chance of acceptance when the process is at the acceptable quality level (AQL) of 1% defective.
   b.   Average outgoing quality limit (AOQL) of 1.5% defective, which is the maximum long-run average if rejected lots are screened 100%.

139. Variables Plan

A statistician is asked to generate a sampling plan for a critical variable x such that the chance of acceptance is 95% at $100p_1 = .1\%$ defective, but only 10% at $100p_2 = .5\%$. A defective is a unit whose x value exceeds the upper spec limit of 65. He gets out Duncan (5), turns to F. E. Grubbs' Poisson distribution table on p. 172, and calculates $n = 1366$ and $c = 3$. In other words, test 1366 units and accept on 3 or fewer defectives. He realizes this large sample size would do nothing to increase his popularity with management! Then he remembers that past histograms for x have been bell-shaped, and the normal distribution fits the data quite well. Furthermore, the s_x control chart has been "in control" with $\bar{s}_x = 5.1$, which can be taken as an estimate of σ_x. What *variables* plan did he recommend?

140. Sequential Plan

This problem was suggested by Dieter Wassmundt, IBM Germany-Mainz. In Problem 139, we considered acceptance sampling plans such that the chance of acceptance is 95% at $100p_1 = .1\%$ defective, but only 10% at $100p_2 = .5\%$. The single-sample attributes plan is $n = 1366$ and $c = 3$, which means inspect 1366 units and accept on 3 or fewer defectives. We saw that a variables plan could reduce the sample size to only 32. Suppose though that the inspection is for a visual defect, so we have no variable to measure. The attributes plan with the least amount of inspection is a sequential plan. Find the plan for the requirement stated here. For acceptance, what is the minimum number of units that must be inspected?

141. Average Sample Size

In the previous problem, suppose the process is at $100p_1 = .1\%$, which is called the acceptable quality level, or AQL. On the average, how many units must we inspect with the sequential plan in order to make the decision to accept or reject?

142. Sample vs. Screen

At a certain visual inspection, a lot of 300 units failed the lot acceptance sampling plan because a random sample of 80 units had 5 defectives, greater than the maximum allowable number of 2. Manufacturing screened the remaining 220 units and found 7 more defectives. Quality Control wonders now if the screening was done properly, since $7/220 = 3.2\%$ is about half the $5/80 = 6.2\%$. Find the probability of this difference or more in the observed direction (Manufacturing lower), if the 220 were inspected with the same care as the 80.

143. Average Run Length

In Problem 7, we saw that if you ask 253 people their birthday, there is a 50–50 chance that at least one of them has your birthday. The 253 can be viewed as the *median* waiting time for your birthday. What is the *mean* waiting time?

144. Average Run Length and Upper Limit

Suppose a manufacturing process shifts to such a high percent defective that the probability of acceptance by the lot acceptance sampling plan is only $P_a = .41$. The run length is defined as the number of lots to rejection. Find the ARL = average run length, and also an upper limit L such that the confidence is 90% or more that the run length is L or less.

145. Percent Defective for Accepted Lots

For a lot of acceptance sampling plan, a random sample of $n = 200$ units from a lot of $N = 8000$ units is submitted to a destructive test. If the sample has $c = 0$ defectives, the lot is accepted. Otherwise, the lot is rejected and scrapped. Sample defectives x for the last 25 lots were: 1, 0, 4, 3, 0, 0, 2, 7, 0, 2, 0, 1, 0, 1, 0, 1, 12, 0, 1, 0, 0, 3, 1, 4, and 1. Estimate the overall percent defective for the accepted lots.

146. Breakeven Fraction Defective

For a certain in-line visual inspection, lots rejected by the sampling plan are screened 100% for defectives. The cost to inspect each unit is $I = \$.10$. Defectives found are scrapped, and the loss per defective is the prior manufacturing cost, $C = \$.50$. On the other hand, if a lot is accepted and processed to the end of the manufacturing line, the loss per defective increases to A = $10.50, which is the fallout cost at final test. Let p be the lot fraction defective at the in-line inspection point. Then for some value of p, the loss must be the same whether we screen the lot 100% or accept it as is. Find this breakeven value p_b for the cost figures here.

147. Sampling Plan for Breakeven Fraction Defective

For the in-line inspection in Problem 146, the lot size N is 15,000 units, for which the Mil-Std-105D (17) sampling tables give sample size code letter M for inspection level II, the usual level. This leads to a sample size of $n = 315$ for normal inspection. Suppose we want an economic sampling plan: one that tends to reject when rejection is cheaper, and tends to accept when acceptance is cheaper. For $n = 315$, what would you recommend for c, the maximum allowable number of defectives in the sample?

148. Fine-Tuning the Sampling Plan for Breakeven Fraction Defective

Suppose you have done Problem 147, settled on $c = 2$, and want to "fine-tune" the sample size. Find n so that $P_a = .50$ at $p_b = .010$.

149. Standard Deviation

A certain product variable x follows a normal distribution within a lot. For the standard deviation, Quality Control wants a lot acceptance sampling plan that operates as follows. If the lot standard deviation is an acceptable $\sigma_1 = 25$, the risk of rejection should be $\alpha = .05$. On the other hand, if the lot standard deviation is an unacceptable $\sigma_2 = 50$, the risk of acceptance should be $\beta = .10$. Find the required sample size n and acceptable sample standard deviation s_a.

150. Mean When Standard Deviation Known

A certain product variable x follows a normal distribution within a lot. For the mean, Quality Control wants a lot acceptance sampling plan that operates as follows. If the lot mean is an acceptable $\mu_1 = .135$, the risk of rejection should be $\alpha = .05$. On the other hand, if the lot mean is an unacceptable .130, the risk of acceptance should be $\beta = .10$. Control chart analysis on the history shows that the within-lot standard deviation is stable at $\sigma_x = .006$. Find the required sample size n and the lowest acceptable sample mean $\bar{x}_a$.

151. Mean When Standard Deviation Unknown

In Problem 150, suppose the within-lot standard deviation is *not* stable but the sample standard deviation s_x averages about .006. Now find the required sample size n and lowest acceptable mean $\bar{x}_a$.

152. Double Sampling Plan

For a certain destructive test, the lot acceptance sampling plan requires an initial sample of 4 units. If all 4 units are good, the lot is accepted. Otherwise, a second sample of 4 units is tested, and the lot is accepted if all 4 units are good. Otherwise, the lot is rejected. What is the probability of acceptance for a lot from a 15% defective process?

153. Double Sampling Plan (*continued*)

For a lot acceptance sampling plan, the acceptable quality level (AQL) is the percent defective for which the probability of acceptance $P_a = .95$, and the lot tolerance percent defective (LTPD) is the percent defective for which $P_a = .10$. Find the AQL and LTPD for the plan in Problem 152.

154. Conventional Double Sampling Plan

In Problems 152 and 153, suppose we make the plan like a conventional double sampling plan by having a reject number after the first sample and basing the decision after the second sample on all units inspected, not just the second sample. With Ac for Accept and Re for Reject, one plan is

Sample	Size	Ac	Re
1st	4	0	2
2nd	4	1	2

Find the AQL and LTPD for this plan.

155. Sample Size for .10 or Less Chance of Acceptance

For a certain visual inspection, a sample of n units is taken from a lot of $N = 120$ units, and the lot is accepted if the sample has $c = 0$ defectives. Suppose the whole lot has $k = 6$ defectives, so it's $6/120 = 5\%$ defective. Find the smallest n such that the probability of acceptance is .10 or less.

156. Total Defectives for .10 or Less Chance of Acceptance

In Problem 155 suppose the sample size is $n = 6$ units. Find the smallest total defectives k such that the probability of acceptance is .10 or less.

CORRELATION

157. Upward Trend?

Manufacturing engineers claim that their continual process improvements have raised the yields over the last 10 days: 71, 68, 70, 74, 70, 69, 71, 74, 73, and 74. Is this upward trend significant at the .05 level?

158. Thank God It's Friday?

Ann Landers recently reported the results of a Mayo Clinic study that gave the number of male deaths by day: Sunday—110, Monday—100, Tuesday—102, Wednesday—97, Thursday—92, Friday—84, and Saturday—119. Are these differences significant at the .05 level?

159. Random Residuals

After finding the least-squares line $y = a + bx$ for some (x,y) points, a statistician wants to check the randomness of the residuals: $y - (a + bx)$. He does this by finding the correlation coefficient r for x and $y - (a + bx)$. For his data, this is .000. What does this tell him about the randomness of the residuals?

160. Intercept/Slope for Least-Squares Line

For three components within a product unit, the least-squares line $y = a + bx$ is calculated for $x = $ component size and $y = $ current. Then the electrical parameter ΔW is calculated as $-a/b$. The x values have no measurement error, but the y values have a $\pm 2\%$ long-term error. However, within the short time that a particular unit is tested, the errors in the y values are all the same. In other words, if one y is, say, 1% high, then they all are. Suppose the (x, y) values are (3, .9475), (6, 2.145), and (20, 8.060). What is the percent measurement error in ΔW?

RELIABILITY

161. Constant Failure Rate?

Fifty-two units are placed on life test for 250 hours. Fails are self-correcting, so the total unit-hours is $52(250) = 13,000$. For the unit-hour intervals 0–1000, 1000–2000, etc., the fails are: 5, 0, 1, 2, 0, 3, 0, 3, 0, 2, 0, 0, and 0. Do you think it would be reasonable to treat the failure rate as constant?

162. Mean Time Between Failures

An electrical tester consists of three repairable components:

    a. Mainframe with mean time between failures (MTBF) = 344 hours
    b. Test can with MTBF = 114 hours
    c. Handler with MTBF = 120 hours

Fails occur at random, and if any component fails, the tester fails. Find the MTBF for the tester.

163. Burn-In

For a certain product unit, history shows that early fails have an exponential distribution for time-to-failure t, with a mean of $\theta = 47$ hours. The other units fail well beyond the guarantee period of one year. How long must the units be burned in to remove 99.9% of the early fails?

164. Tool Fixed?

A manufacturing tool has a history of random fails, with a fail every 20 hours, on the average. This is considered unacceptable, so the tool gets some special maintenance work. After the work, how long must the tool run with no fails in order to be 90% confident that there has been some improvement?

165. 1% Fail Point for Lognormal Distribution

For a certain product unit, the time-to-failure t follows a lognormal distribution, which means that ln t follows a normal distribution. Suppose that the mean is $\mu_{\ln t} = 8.709$ and the standard deviation is $\sigma_{\ln t} = .312$. Find $t_{.01}$, which is the t value for which the cumulative probability is .01. In other words, 99% of the lifetimes will exceed $t_{.01}$.

166. A Bad Connection

At the 10-hour module burn-in, a fraction f of the sockets have a connection problem such that the probability of a bad connection is p. For the other fraction $1 - f$, the probability of a bad connection is zero. Engineering wonders what the effect would be of unplugging and replugging all the modules at 5 hours. Find the mean burn-in time per module for each of these two methods: No Replug and Replug.

167. Zero Fails

A sample of 350 product units has zero fails after 500 hours of stress. If we make no assumption about the time-to-failure distribution, what is the upper 90% confidence limit for the cumulative percent failing by 500 hours?

168. Zero Fails and Constant Failure Rate

In Problem 167, suppose we assume a constant failure rate. Now what is the upper 90% confidence limit for the cumulative percent failing by 500 hours?

169. Zero Fails and Constant Failure Rate (*continued*)

In Problems 167 and 168, suppose for the product use end-of-life is only 250 hours. Then we want to know the upper 90% confidence limit for the cumulative percent failing by 250 hours. For no assumption about the time-to-failure distribution, the limit is the same as it was in Problem 167 because it's still 0 fails in 350 units. What is the new limit under the constant failure rate assumption?

170. Spare Parts

History shows that a critical part for a manufacturing tool fails at random, with a constant failure rate of .35 fails per month. Spare parts are ordered once a month. How many parts should you have on hand at the start of a month in order to be at least 99% confident that you'll have enough parts to get you through the month?

MOVING AVERAGES

171. Finding the Total from Moving Sums

At reliability test, a sample of 100 units is tested each week, and the moving sums of defectives for 13 weeks are calculated. For a certain calendar year, the sums for weeks 1–13, 2–14, 3–15, . . . , 40–52 were (left to right, row by row):

```
51  50  52  51  51  46  46  48  48  56
55  52  52  50  54  55  59  59  65  69
69  67  60  59  60  58  58  53  51  47
46  39  37  35  37  37  37  38  42  43
```

Now Engineering wants to know the sample percent defective for the whole year. What is it?

172. Trend Analysis on Moving Averages or Sums

In the previous problem, Engineering also wants to know why the moving sum got worse from week 20 to week 30, and then improved after that.

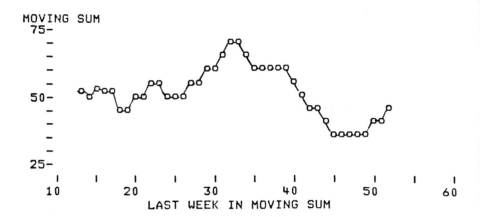

MOVING SUM

LAST WEEK IN MOVING SUM

What do you think?

173. Finding Individuals from Moving Sums

Suppose we have only the moving sums from Problem 171. Derive the defectives for each of the 52 individual weeks.

SIGNIFICANCE TESTS

174. 0/n vs. n/n

Suppose method 1 makes 0% defective units, and method 2 makes 100% defective units. How large a sample size with each method is required to get a significant difference with Fisher's Exact Test for two proportions? Use a one-tail test at the $\alpha = .05$ significance level.

175. 0/n vs. 1/1

A tool has produced n consecutive defective widgets, one at a time. An adjustment is made to the tool, and the next one is nondefective. How large must n be for significance at the .05 level?

176. x_1/n_1 vs. 0/n_2

A manufacturing process has produced $x_1 = 5$ defectives in $n_1 = 50$ units. After modifications are made to the process, there are

$x_2 = 0$ defectives in n_2 units. How large must n_2 be for a significant difference at the .05 level?

177. x_1/n_1 vs. x_2/n_2

At a certain point in the manufacturing line, two methods give the following fractions defective for random samples of units: control = 4/200, experimental = 0/200. What is the probability of such a large difference in this direction just due to chance, if the methods really have the same fraction defective in the long run?

178. Old Process Percent Defective vs. New

A new process has made six lots, and a random sample of 160 units from each lot has been inspected. Defectives by sample are 4, 7, 5, 17, 9, and 1, for an overall percent defective of 100(43/960), or 4.5%. The old process had been averaging 5.9% defective. What is the probability of such a good performance by the new process, if it's really the same old 5.9% in the long run?

179. Fails for Four Tools

For the same operating time, four manufacturing tools have fails of 9, 4, 5, and 6. The engineer wonders if the tool reliabilities are really different. Are the differences in the observed fails significant at the .05 level?

180. Picking the Better Mean

A manufacturing manager wishes to run an experiment to pick the better of two electrode pastes with regard to adhesion to the substrate. He doesn't believe in significance tests because any sample difference, no matter how small, is significant if the sample size is big enough. Further, he rejects the hypothesis of no difference *before* the experiment because he knows there has to be *some* difference between the two pastes. The manager decides to run a sample of $n = 20$ substrates with each paste type, and simply pick the paste with the higher sample mean as being better. History shows that $x = $ adhesion follows a normal distribution with a standard deviation of $\sigma_x = 74$, regardless of paste type. A difference in population means of 30 would be important. If this is really the case, what is the probability that the better paste is picked?

181. Picking the Better Standard Deviation

To measure a certain dimension x, a manufacturing engineer must buy a tester from one of two rival companies. He wants the tester with the better repeatability. To make this decision, he decides to take n measurements with each tester on the same product unit, and then buy the tester with the lower sample standard deviation:

$$s_x = \sqrt{\frac{\Sigma(x - \bar{x})^2}{n - 1}}$$

If the ratio of the smaller population standard deviation to the larger is .75 or less, he wants to be at least 90% confident of making a correct selection. A ratio between .75 and 1.00 would not be an important difference to him, and he would be happy with either tester in that case. How large should n be?

182. Picking the Lower Percent Defective

A certain manufacturing method has been giving 2% defective, and Engineering wants to compare it with a new method which they hope is 1% defective or better. They decide to make 480 units with each method and then pick the method with the lower sample percent defective. A tie will be settled by some random procedure, such as a coin toss. If the population percents defective are really the hypothesized values of 2% and 1%, what is the probability that the new method will be picked?

183. Comparing Machines

Hours of down-time for four machines were as shown.

	Machine			
Month	A	B	C	D
Jan.	7.8	8.1	6.0	7.2
Feb.	8.0	7.2	8.3	6.9
March	7.7	7.8	7.5	7.1
April	8.2	7.6	7.3	7.9
May	8.0	7.3	7.6	8.1
Avg.	7.9	7.6	7.3	7.4

Are the differences in the machine averages significant at the .05 level?

184. Fisher's Test?

A unit has two components, A and B. Engineering suspects that due to poor design A's quality is worse than B's. Test results for 124 units are as shown.

	A		
B	Bad	Good	All
Bad	6	1	7
Good	7	110	117
All	13	111	124

Hence, A had $13/124 = 10.5\%$ bad, and B had $7/124 = 5.6\%$. Is this difference significant at the .05 level?

185. Inspectors/Several Units Inspected Once

The same 69 units are inspected visually by four inspectors, and the results with 1 for bad and 0 for good are as shown.

		Inspector		
Units	A	B	C	D
4	1	1	1	1
2	1	1	0	1
3	0	1	1	1
1	0	1	0	1
59	0	0	0	0
69	6	10	7	10
% Bad	9	14	10	14

This means, for example, that inspector A saw 6 defectives: 4 from the first row in the table and 2 from the second. Are the differences in inspectors significant at the .05 level?

186. Inspectors/One Unit Inspected Several Times

Quality Control wants to compare three automatic visual inspection tools to see if they give the same results, on the average. The same product unit is inspected time after time by the three tools, and results are as shown.

Tool	Defects at a time											
1	2	1	4	3	3	2	0	1				
2	1	3	2	2	2	0	3	1	2	1	2	1
3	1	3	2	0	2	2	4					

All defects called by the tools are true defects—that is, there are no false defects. Are the differences in tools significant at the .05 level?

187. Before vs. After

Engineers feel a certain process step is unnecessary, so the step is eliminated. Inspection results before and after the change, for samples of $n = 30$ units per lot, are as shown.

	Lots	
x = Good units	Before	After
30	19	1
29	10	5
28	2	0
27	4	2
26	2	1
25	0	0
24	1	0

Has the quality degraded? Use a one-tail test for a significant decrease in the mean x/n at the .05 level.

188. Suicide Prevention Test

A recent newspaper article discussed a suicide prevention test that measures a person's feeling of hopelessness. A total of 207 patients suffering from depression were given the test, and then they were tracked over the next 10 years. Of the 14 patients who committed suicide, 13 scored high on the hopelessness scale. Forty others who did not commit suicide also scored high. The psychiatrists claim this shows that people who score high have more suicides. Do you agree?

189. Within-Lot Variability

For a check on the uniformity of product from a vendor, samples from six lots are measured for a critical electrical parameter X.

	Lot					
Unit	1	2	3	4	5	6
1	16.5	15.7	17.3	16.9	15.5	13.5
2	17.2	17.6	15.8	15.8	16.6	14.5
3	16.6	16.3	16.8	16.9	15.9	16.0
4	15.0	14.6	17.2	16.8	16.5	15.9
5	14.4	14.9	16.2	16.6	16.1	13.7
6	16.5	15.2	16.9	16.0	16.2	15.2
7	15.5	16.1	14.9	16.6	15.7	15.9

Can the differences in the within-sample variability be reasonably attributed just to chance?

190. Lot Means

Now look at the sample means in Problem 189. Can *these* differences be reasonably attributed just to chance? Use a .05 significance level.

191. Sample vs. Rest of Lot

Plant A inspects a random sample of $n = 32$ modules from a lot of $N = 250$ modules and finds $x = 1$ defectives. This passes the sampling plan, so the defective is removed, and the other 249 modules are shipped to Plant B, which finds 15 defective modules when the 249 are plugged into cards. Plant B thinks this is too many, since $15/249 = 6.0\%$ is twice as big as $1/32 = 3.1\%$. They claim that Plant A did not fairly inspect a random sample from the lot. Do they have a case?

CONFIDENCE INTERVALS

192. Confidence Interval for New Process Percent Defective

A new process has made six lots, and a random sample of 160 units from each lot has been inspected. Defectives by sample are 4, 7, 5, 17, 9, and 1, for an overall percent defective of 100(43/960), or 4.5%. Find the 95% confidence interval for the long-run, overall percent defective from this process.

193. Split Lots

Ten split lots were run to compare an experimental method with the control method. Control yields by lot were:
 86.5   87.8   85.1   82.4   93.2   82.4   82.4   82.4   78.4   86.5
Experimental yields in the same order were:
 86.5   91.9   93.2   86.5   95.9   90.5   85.1   90.5   85.1   94.6
Find the 95% confidence interval for the long-run mean difference in yields, experimental minus control.

194. Making a Good Widget

A process produces special-order widgets one at a time, with a random 53% of them defective. How many widgets must be started to have a 90% confidence that at least one is good?

195. Winning at Solitaire

While recuperating from an appendectomy, a certain statistician plays 50 games of solitaire and wins 18 times. He wonders what his winning percent would be in the long run. Find the 95% confidence interval for this.

196. Total Units in Stock

There is some question about the total number of units in stock, since accurate records have not been kept. For an estimate, $n = 10$ lots are randomly selected from the total of $N = 100$ lots, and a count x is made for each lot. The mean and standard deviation are $\bar{x} = 4851.80$ and $s_x = 658.76$. Find 95% confidence limits for the total number of units in stock.

197. Distribution-Free Tolerance Limits

After a process change, a random sample of 100 product units is measured for a variable x. The lowest value is 4.2, and the highest is 15.0. With 95% confidence, at least what percent of the population is between 4.2 and 15.0?

198. Normal-Distribution Tolerance Limits

A histogram of the sample values in the previous problem shows that the distribution is approximately bell-shaped with $\bar{x} = 9.8$ and $s_x = 2.2$. Use a normal distribution assumption to find limits for x such that the confidence is 95% that at least 99% of the population is within the limits.

199. Normal-Distribution Tolerance Limits (*continued*)

In the previous problem, suppose we want to find limits for x such that the confidence is 95% that at least 95% of the population is within the limits. What are the limits?

200. Distribution-Free Confidence Limits for Percent Defective

In Problems 197–199, we saw some x measurements after a process change. The design people have now decided that in order for the product unit to work properly, the x should be greater than or equal to $L = 6.0$. For the random sample of 100 units measured, 2 were below 6.0. Without making any assumptions about the form of the x distribution, find the estimate for the population percent defective, and also the upper limit with 90% confidence.

201. Normal-Distribution Confidence Limits for Percent Defective

In the previous problem, suppose we want to use the normal distribution assumption, since the histogram of x values is approximately bell-shaped with $\bar{x} = 9.8$ and $s_x = 2.2$. Now find the estimate for the population percent defective, and the upper limit with 90% confidence.

202. Election Night

With 41% of the vote counted on election night, the results for lieutenant governor are as shown.

Candidate	Votes
Dean	29,264
Auld	24,723
Hopps	876
Simons	868

If no candidate receives more than 50% of the vote, then the legislature decides. With these results in hand, Auld concedes to

Dean. Was this a good decision? Assume the votes can be treated
as a random sample from all the votes.

203. Two-Stage Sample for Percent Defective

Product units called chips are made in wafer form, with $M = 100$
chips/wafer. From a particular lot of $N = 75$ wafers, a random
sample of $n = 5$ wafers is taken, and then a random sample of
$m = 10$ chips from each of these wafers is inspected. Defective
chips by wafer are: 5, 2, 6, 9, and 2. Find the 95% confidence
interval for the lot percent defective.

204. Tool Set-Up

A manufacturing tool makes a product unit with a critical dimension x. At set-up, the operator measures x for a sample of n
units to see if the tool is centered at the nominal value of 24.91.
History shows that the standard deviation during this set-up period is $\sigma_x = .82$. How large must n be in order to have 95%
confidence that the error in the sample mean $\bar{x}$ is less than $\pm.5$?

205. Repeatability Error

In order to check the repeatability for a new measurement tool,
Quality Control measures the same unit $n = 20$ times over a two-
week period. The standard deviation for the x measurements is
$s_x = .48$. Suppose we take $3s_x = 1.44$ as the *estimate* for the
population $3\sigma_x$ over an indefinitely large number of readings.
Since nearly all the population readings will be within three standard deviations from the mean, the unknown $3\sigma_x$ can be considered
the maximum repeatability error for the tester. What is the upper
95% confidence limit for the new measurement tool's $3\sigma_x$?

206. Repeatability Error (*continued*)

In Problem 205, suppose we want the upper 95% confidence limit
for $3\sigma_x$ to be only 10% above $3s_x$. How large must n be?

207. Defects in Windows

The total surface of a product unit is $T = 100$ square centimeters. For visual inspection, the area is divided into windows of size $A = 2$ square centimeters. Then a random sample of $n = 20$ windows is inspected to get a defect count x by window. The reported value is total defects over total area inspected, which is the defect density as defects per square centimeter: $(\Sigma x)/(nA)$. For a particular unit, the x counts were 0, 0, 2, 1, 0, 0, 6, 1, 7, 1, 0, 8, 2, 4, 1, 1, 0, 3, 0, and 0. Find the 95% confidence interval for the true defect density for the unit. Let Student's t be 2.

208. Defects in Windows (*continued*)

In Problem 207, suppose the window area is changed to kA, but the new sample size is n/k, so the total area inspected is the same: $(n/k)(kA) = nA$. If $k = .25$, and the unit were reinspected, can you use the Problem 207 data to estimate what the new data would give for the 95% confidence interval for the true defect density for the unit, *without* seeing the new data?

209. Voids in Lines

Suppose voids in aluminum lines occur at random with a constant void density, which is estimated as $\lambda = .3$ voids per unit length. We wish to inspect a length L to characterize the percent aluminum area voided. Find L such that there is 90% confidence of seeing at least 10 voids?

210. Voids in Lines (*continued*)

In Problem 209, the length L was spread over a random sample of $n = 5$ equal-size lines from an indefinitely large population of lines. By line, the values for the percent aluminum area voided were .8872, 1.3308, 1.0259, 1.2199, and 1.1922. Find the 95% confidence interval for the overall percent aluminum area voided in the population of lines.

NONSTATISTICAL PROBLEMS

211. Three Navaho Women

Three Navaho women sit side by side on the ground. The first woman, who is sitting on a goatskin, has a son who weighs 140 pounds. The second woman, who is sitting on a deerskin, has a son who weighs 160 pounds. The third woman, who is sitting on a hippopotamus skin, herself weighs 300 pounds. This suggests a variation on a famous geometric theorem. What is it?

212. Switched Labels

You work in the back room of a grocery store, where three crates of fruit have just come in with these labels: "apples," "oranges," and "apples and oranges." The boss tells you he has received a phone call reporting that all three crates are mislabeled! You are in a hurry, so you want to pick a piece of fruit from only one crate and then correctly switch the labels. How do you do it?

213. Triangle Former

A triangle former picks two points at random on a line segment. What is the probability that the three segments determined can form a triangle?

214. Potato Race

As a fun game at his son's birthday party, a farmer proposed a potato race. He had previously placed 40 piles of potatoes, one potato per boy in each pile, along the road. The distance to the first pile was only one yard, then three yards between the first and second piles, five yards between the second and third, etc., with the distance between piles increasing by two yards for each new pile. The race was to see who could be the first one to bring back a potato from each pile, one at a time—that is, a potato from the first pile, then out again for one from the second pile, etc. How long was the race?

215. Tester Availability

An electrical tester consists of four independent tools that operate in parallel, so if a tool goes down for repairs, the others keep testing. In the long run, each tool is down 10% of the time for random failure. As an example, for a clock time of 3 hours, the *scheduled* test time would be $3(4) = 12$ hours, but due to failures the *actual* test time might be only 9.6 hours. Find the long-run tester *availability*, which is the actual test time as a percent of the scheduled test time.

216. Math Test

A statistician gave math tests to everyone who lived in a village of 6000 people and at the same time measured the length of their feet. He found a strong positive correlation between foot length and math ability. Explain.

217. Crossing a Moat

You are a knight in shining armor who must rescue a fair damsel held captive in a castle on land surrounded by a moat. All you have to cross the moat are two boards, each 9.9′ long. How do you do it?

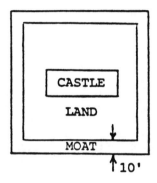

218. Selling Sheep

Two brothers sell their flock of sheep. By chance, the number of dollars received for each sheep turns out to be the same as the total number of sheep. The money is in $10 bills, and $1 bills totaling less than $10. They try to divide the bills equally, but the number of $10 bills is odd, so they can't. To solve the problem, one brother takes the extra $10 bill, lets his brother take the $1 bills, and then writes him a check so the split is equal. How much is the check?

219. A Boy, a Girl, and a Dog

A boy is out walking his dog when he sees his girlfriend walking toward him a quarter-mile away. The dog sees her too, and starts running toward the girl. The boy and girl continue walking toward each other at 3 miles per hour, and the dog runs first to the girl, then back to the boy, etc., at 12 miles per hour. How far does the dog run before all three meet?

220. Bookkeeper

At work one day, a bookkeeper for a toy company noticed that the word "balloon" had two sets of double letters, one after the other. "I wonder," he thought, "if there is a word containing *three* sets of double letters, one right after the other." Can you think of one?

221. Two Coins

I have two coins in my pocket. Together they total $.30. One of them is not a nickel. What are the coins?

222. Shoeing a Horse

A man takes his horse to the blacksmith for four new shoes. The blacksmith wants $50 for the job, but the man complains that this is too much. "O.K.," says the blacksmith, "I'll just charge you 1¢ for the first nail, 2¢ for the second, 4¢ for the third, and so on. We'll just double the price for each new nail. There are only 32 nails in all." The man readily agrees to this new deal. How much does he pay for the job?

223. Foot Race

A boy and girl run a 100-yard race, and the boy wins by 5 yards. To make their next race even, the boy offers to start 5 yards behind the starting line. The girl accepts the offer. If they run at the same speed as before, who wins, or is it a tie?

224. Average Speed

A man drives his car 1 mile to the top of a mountain at 15 miles per hour. How fast must he drive the 1 mile down the other side to average 30 miles per hour for the whole trip of 2 miles?

225. Poles With Guidewires

Two poles are connected by guidewires as shown. Find x, the distance between the poles.

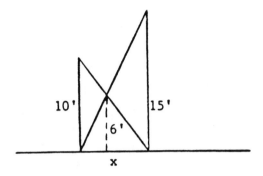

226. Calculator Trick

Let x be any positive non-zero number. Take the square root of x and double it. Then treat your answer the same way: Take the square root and double it. If you continue this indefinitely, what is the limiting value?

227. Another Calculator Trick

Let a and b be any two positive non-zero numbers, and suppose they are the first two numbers in a sequence. To get the next number, add 1 to b and divide by a. Continue this procedure indefinitely to extend the sequence: add 1 to the last number and divide by the number before that. What is the limiting value?

228. Half-Full

Pat and Mike are having their last glass of beer before hitting the road. Pat finishes his and gets up, but Mike says, "Sit down. I'm not even half-done." Pat picks up Mike's cylindrical glass, inspects it, and says, "Yes, you are." How can Pat tell so easily that Mike's glass is less than half-full?

229. Dog, Goose, and Corn

You want to cross a river with your dog, goose, and corn, but your boat is so small that you can take only one thing across at a time. That isn't the only problem. If you leave the dog and goose alone, the dog will eat the goose, and if you leave the goose and corn alone, the goose will eat the corn. How do you get everything across?

230. Missionaries and Cannibals

Here is another river-crossing problem. Three missionaries and three cannibals must cross a river in a boat that can hold only two people. The cannibals do not need to be guarded, but at no time can the cannibals outnumber the missionaries in a group, for obvious reasons. How do they get across?

231. Picking Grapes

A farmer has just finished a hard day's work in the field. Before he returns to the house, he would like to pick a bunch of grapes from a vine that borders his property. In the following figure, F is the farmer, G is an arbitrary point along the grapevine, H is the house, and distances are in yards. Find the x that minimizes the distance the farmer must walk.

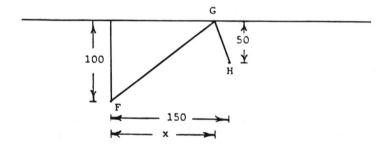

232. Spaghetti Measure

When you cook spaghetti, you make either too much or not enough, so you decide to make a spaghetti measure. This is a piece of wood with a hole to stick the spaghetti through for one serving, another size hole for two servings, etc. You decide that for one serving, a 7/8-inch diameter is about right. What should the diameter be for *i* people?

233. Interesting Circles

Two circles intersect such that the radii shown form a right angle. Find the difference in the nonoverlapping, unshaded areas.

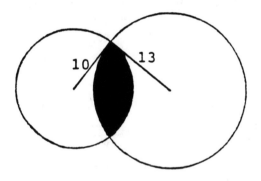

234. Man in a Coffin

This one is from Ray Gonda, IBM Burlington. At a funeral, Frank points to the man in the coffin and mysteriously says, "Brothers and sisters have I none, but that man's father is my father's son." What is the relationship between Frank and the man in the coffin?

235. A Circular Swimming Pool

A wealthy executive has a circular swimming pool that is 50 feet in diameter. For her morning dip, she starts from point A on the edge of the pool and swims across the center of the pool to point B on the edge. Then she turns 150° to her left and swims to point

C on the edge. At what angle must she now turn in order to return
to point A?

236. Frog in a Well

A frog is at the bottom of a 40-foot well. Every day he jumps up
3 feet but then falls back 2 feet. How many days does it take him
to get out of the well?

237. Water in Buckets

You have a 10-gallon bucket full of water, an empty 7-gallon
bucket, and an empty 3-gallon bucket. How do you get 5 gallons
of water in one bucket? You want to dump it on your spouse's
head to celebrate your fifth wedding anniversary.

238. Water in Buckets (*continued*)

The solution to Problem 237 required 9 steps. Find an 8-step
solution.

239. Doubling Your Money

If a principal of *P* dollars is deposited in a bank that pays interest
at an annual rate of 100*i*% compounded annually, then *P* will
double in *n* years. Economists like to express *n* as $D/(100i)$. Find
D for 100*i* equal to 1, 8, and 15.

240. Transportation Problem

You want to take your motorcycle into the shop for some repair
work, but you can't find anyone to pick you up at the shop after
you drive it there, or take you there to pick it up after the work
is done. However, you do have a 10-speed bike and a car with a
bike rack. How can you solve the transportation problem without
any outside help?

241. Summing Integers

According to legend, when the famous mathematician Gauss was
10 years old, his class at school was asked to add up the integers

from 1 to 100. Gauss quickly wrote down the answer without adding up the 100 numbers. What is the answer and how did he find it?

242. Summing Squares of Integers

Find the sum of the *squares* of the integers from 1 to n.

243. Summing Cubes of Integers

This one is from Lyle Dockendorf, IBM Rochester. A king has an unmarried daughter pursued by two suitors. Taking a cue from the 64 squares on a chessboard, suitor A offers the king this many pieces of silver for the daughter's hand in marriage:

$$(1 + 2 + 3 + \ldots + 64)^2$$

Suitor B counters with an offer of this many pieces of silver:

$$1 + 2^3 + 3^3 + \ldots + 64^3$$

Find the sum for each of these offers.

244. Speed of a Pitch

When Bob Feller pitched for the Cleveland Indians, they didn't have radar guns, so they estimated the speed of his pitch by having him throw the ball the 60' 6" to the plate at the same time that a nearby motorcycle was following a parallel route at 100 miles per hour. Ball and motorcycle finished in a tie, so Feller's speed was estimated as 100 miles per hour. With today's digital stopwatch and pocket calculator, how could you estimate the speed of a pitch?

245. Send More Money

Substitute digits for the letters so the addition is correct:

$$\begin{array}{c}
S\ E\ N\ D \\
\underline{M\ O\ R\ E} \\
M\ O\ N\ E\ Y
\end{array}$$

246. Triangle Area

Find the area of this triangle:

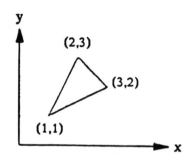

247. Average Time to Get a Good Unit

Suppose a manufacturing process takes 2 hours, and the yield is 40% good units. Rework of a bad unit takes the same time and has the same yield. What is the average time to get a good unit?

248. Rope Around the Earth

One day you start walking while trailing a long rope behind you. You don't stop until you get back to where you started, so the rope makes a circle around the Earth's center with a diameter of 7913 miles. Then you let out 3 more feet of rope and make another circle around the Earth. How high off the ground is the rope?

249. Rope Around the Earth (*continued*)

Now suppose you take the rope in Problem 248, with the 3 feet added, and pull it straight up away from the Earth's center until there is no slack. How high off the ground is the high point of the rope? Find the answer to the nearest foot.

250. Buying New Socks

While you are waiting for your wash to get done at the laundromat, you think about buying a new set of socks for the week because yours are getting pretty thin in spots. If you do your wash once a week, how many pairs must you buy to get you through the week with clean socks?

251. Chip Height

A rectangular chip with pads underneath sits on a substrate. If the pads are not the same size, the chip will tip. A critical dimension is chip height, which is the distance from the substrate to the lowest point on the chip. If we assume the chip and substrate are planes, then we can measure the height at the four corners, and the lowest height is the required dimension. Actually, if we have three corner heights, the height at the fourth corner is determined. Let z_1, z_2, and z_3 be the three measured corner heights, with z_1 and z_3 opposite each other. Find the height z_4 at the fourth corner.

SOLUTIONS

1. TV Game Show

Yes, he should swap, for a 2/3 probability of winning, instead of 1/3. After the contestant picks up a box, the probability must be 1/3 that he has the $100 box, and 2/3 that it is on the table. The host's showing an empty box on the tables does not change these probabilities, because he always does this. At least one of the boxes on the table *has* to be empty, and he knows the contents. This is the key. The host's selection is not a random one. If it were, certainly the probabilities would be 1/2 for the two remaining boxes. Thanks to the host, the contestant can now "take the table." If the $100 is there, he's got it.

2. Chemical Spill

75%. The probability of at least one spill is 100% minus the probability of none. With $p = 4/1036$ for the fractional probability of a spill in one transfer, the probability of *no* spills in the 358 is

$(1 - p)^{358} = .25$, or 25%. Subtracting this from 100% gives the answer.

3. Flat Tire

1/4. The chance that a particular boy picks, say, the left front tire is 1/4; the chance they *both* pick that tire is $(1/4)^2$. This is also the probability for each of the other tires, so summing the probabilities for these mutually exclusive events gives the required answer: $4(1/4)^2 = 1/4$.

4. Duplicate Birthdays

23. Rather than calculate the probability of duplicates directly, it's easier to come in the "back door" and calculate the probability of *no* duplicates. This subtracted from one gives the probability of duplicates. For no duplicates, the first person can have any of the 365 birthdays, the second any of the remaining 364, the third any of the remaining 363, etc. The probability of duplicates then is:

$$1 - \left(\frac{365}{365}\right)\left(\frac{364}{365}\right)\left(\frac{363}{365}\right) \dots \text{ to } n \text{ terms}$$

The product can be evaluated on a calculator as $364 \div 365 \times 363 \div 365 \dots$ For $n = 22, 23,$ and 24, the probabilities of duplicates are .476, .507, and .538.

5. Birthday Problem Variation

7. It is unfortunate that the months do not all have the same number of days d. However, the true probabilities must lie between those for $d = 30$ and those for $d = 31$. The probability of duplicates is one minus the probability of *no* duplicates. For no duplicates and $d = 31$, the first person can have any of the 31 days, the second any of the remaining 30, the third any of the remaining 29, etc. The probability of duplicates then is

$$1 - \left(\frac{31}{31}\right)\left(\frac{30}{31}\right)\left(\frac{29}{31}\right) \dots \text{ to } n \text{ terms}$$

For $n = 6$ and 7, the probabilities are .403 and .518. With $d = 30$, there is little change: .414 and .531, so the answer is $n = 7$.

6. Getting off an Elevator

.06. Did you recognize this as the classical birthday problem in disguise? We are now asking the floor number instead of the day born. For no duplicate floors, the first person can have any of the 10 floors, the second any of the remaining 9, the third any of the remaining 8, etc. The probability of no duplicates then is:

$$\frac{10}{10} \left(\frac{9}{10}\right) \left(\frac{8}{10}\right) \cdots \left(\frac{4}{10}\right) = .06$$

Although seemingly rare, the event fails be significant at the .05 level.

7. Your Birthday

253. The probability that at least one person has your birthday is one minus the probability that no one does, so we must have

$$1 - \left(\frac{364}{365}\right)^n = .50 \Rightarrow \left(\frac{364}{365}\right)^n = .50$$

Taking the logs of both sides of this equation and solving for n gives 253.

8. General Birthday Problem

$-\ln(1 - C)$. The probability of at least one success is one minus the probability of none, so we have

$$1 - \left(\frac{m - 1}{m}\right)^n = C$$

$$n = \frac{\ln(1 - C)}{\ln\left(1 - \dfrac{1}{m}\right)}$$

The $\ln[1 - (1/m)]$ can be expressed as an infinite series:

$$\ln\left(1 - \frac{1}{m}\right) = -\left[\left(\frac{1}{m}\right) + \frac{\left(\frac{1}{m}\right)^2}{2} + \frac{\left(\frac{1}{m}\right)^3}{3} + \cdots\right]$$

This is approximately $-1/m$ for large m, and this substitution leads to

$$\frac{n}{m} \simeq -\ln(1 - C)$$

For example, if $C = .90$, then $n/m \simeq 2.3$. If $n/m = 1$, $C \simeq .63$.

9. A Famous French Statistician

Poisson. We have

$$e^{-n/m} \simeq 1 - C$$

The $1 - C$ is the probability of zero successes in the n trials. Remember that for a Poisson distribution with mean number of successes λ, the probability of zero successes is $e^{-\lambda}$. Here the n/m is the mean number of successes, since we actually have a binomial distribution model with n trials, $p = 1/m$ probability of success on any one trial, and mean number of successes $\lambda = np = n/m$. Hence, the Problem 8 solution is the famous Poisson approximation to the binomial. The infinite series is the basis for the approximation See Feller (6). The usual rule is that the approximation is satisfactory if $p < .10$. Here $p = 1/m$, so the requirement is $1/m < .10$, or $m > 10$. If $m = 10$ and $C = .90$, then the approximate n is $-10(\ln .10) = 23$, and the exact n is $(\ln .10)/(\ln .90) = 22$.

10. Finding a Bad Component on a Board

C. This is a finite population version of Problem 8. Now we must allow for depletion. The probability of finding the bad one is one minus the probability of not finding it:

$$1 - \left[\frac{m-1}{m} \cdot \frac{m-2}{m-1} \cdot \frac{m-3}{m-2} \cdots \frac{m-n}{m-n+1} \right] = C$$

$$\frac{n}{m} = C$$

For an intuitive solution, imagine a dartboard divided into m equal areas, and any n of these areas are red. Think of a dart as the bad component. If we put on a blindfold and throw the dart at the board, the probability that the dart lands on red is n/m.

11. Finding a Bad Circuit in a Component

990. The toughest defective component to find is one with just one bad circuit, and we must have a .99 probability of finding it in the n. The probability of finding it is one minus the probability of not finding it:

$$1 - \left(\frac{N-1}{N} \cdot \frac{N-2}{N-1} \cdot \frac{N-3}{N-2} \cdots \frac{N-n}{N-n+1} \right) = \frac{n}{N}$$

Setting $n/N = .99$ and replacing N with 1000 gives $n = 990$. In general, the percent confidence of finding the one bad circuit is the same as the percent of the total circuits tested.

12. Chips on Modules

16.7%. The probability of at least one defective chip on a module is one minus the probability of none:

$$1 - (1 - .03)^6 = .167, \text{ or } 16.7\%$$

This value is quite a bit higher than the 3% for chips. In general, components in a system must have a very low percent defective if the system is to have a respectably low percent defective.

13. Passing Final Test

.092. Think of the defects as bullets. Then the probability of passing final test is the probability of dodging all the bullets:

$$(1 - .10)^1(1 - .20)^3(1 - .80)^1 = .092$$

14. Leaving a Row

2/45, or .04. The probability that the first finisher has no problem is the probability that he is on one end of the row, and this is 2/6. Then the probability for the second finisher is 2/5, etc. The overall probability of no problem is the product:

$$\frac{2}{6} \cdot \frac{2}{5} \cdot \frac{2}{4} \cdot \frac{2}{3} \cdot \frac{2}{2} \cdot \frac{1}{1} = \frac{2}{45}$$

15. A Man and a Woman

11/36. Let x be the minutes after five o'clock when the man arrives, and y be the same thing for the woman. In order for the man and woman to be in the bar at the same time, x and y must be within 10 minutes of each other, or $y \geq x - 10$ *and* $y \leq x + 10$. This is the shaded area in the figure. All points within the square are equally likely, so the required probability is the ratio of the shaded area to the total. Since the shaded area is the total

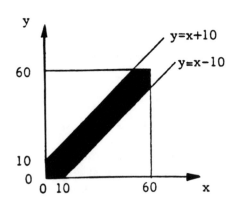

minus the two unshaded triangles, we have

$$\frac{3600 - 2\left(\frac{1}{2}\right)(50)^2}{3600} = \frac{11}{36}$$

16. A Man and a Woman Waiting

18 minutes. In the problem 15 solution, replace 10 with the required t, and the probability of being in the bar at the same time becomes

$$\frac{3600 - 2\left(\frac{1}{2}\right)(60 - t)^2}{3600} = \frac{120t - t^2}{3600}$$

Setting this equal to .5 and solving for t with the well-known quadratic formula from algebra gives $t = 18$ minutes.

17. Door Prizes

1/120. For any one of the equally likely sets of 5 tickets, there are $5! = 120$ possible orders, only one of which is ascending. Interestingly, the 1/120 answer does not depend on the total number of tickets.

18. Putting Letters in Envelopes

.63. See Feller (6). The probability of at least one match for N letters and N envelopes is found by taking the reciprocals of the factorials from 1 to N, and then summing them with alternating signs starting with a $+$:

$$\frac{1}{1!} - \frac{1}{2!} + \frac{1}{3!} - \ldots \pm \frac{1}{N!}$$

This gives .50 for $N = 2$, .67 for $N = 3$, .62 for $N = 4$, and .63 for $N \geq 5$.

19. Two Children

1/3. With B for boy and G for girl, there are four equally likely orders of birth for two children: BB, BG, GB, and GG. Since we know there is at least one boy, we can rule out GG. This leaves three equally likely possibilities, of which only one is both boys.

20. Two Clocks

1/2. When the time was set on the new clock, the old-clock minutes had changed x seconds ago, with x equally likely to be anything from zero to 60. Then the clocks agreed for $60 - x$ seconds until the old clock changed, disagreed for x seconds until the new clock changed, etc. The joint probability of a particular x and then disagreement later at a random point in time is $(dx/60)$ $(x/60)$. If we integrate (sum) this over all possible x, we get the required probability:

$$\frac{1}{3600} \int_0^{60} x \, dx = \frac{1}{3600} \left(\frac{x^2}{2} \right) \Big]_0^{60} = \frac{1}{2}$$

21. Three Vendors

67%. This calls for Bayes' Theorem, named after the English clergyman Thomas Bayes (1702–1761). The problem is sometimes given using urns instead of vendors, and black and white balls instead of defectives and nondefectives. The problem is then, "Given that a random ball from a random urn is black, find the probability it came from urn A." Here suppose there are N total components. Then the required fraction is the ratio of the ways to get a defective from Vendor A to the total ways to get a defective:

$$\frac{.40(.10)N}{.40(.10)N + .35(.05)N + .25(.01)N} = \frac{2}{3}, \text{ or } 67\%$$

For the urns/balls problem, N would be an indefinitely large number of draws, and the .40, .35, and .25 would all be replaced by 1/3, for an answer of 62%.

22. Changing Partners in a Card Game

One. With A, B, C, and D for the players, let the three games be AB vs. CD, AC vs. BD, and AD vs. BC. Possible outcomes are:

	Winner by game			Games won			
Outcome	1	2	3	A	B	C	D
1	AB	AC	AD	3	1	1	1
2	AB	AC	BC	2	2	2	0
3	AB	BD	AD	2	2	0	2
4	AB	BD	BC	1	3	1	1
5	CD	AC	AD	2	0	2	2
6	CD	AC	BC	1	1	3	1
7	CD	BD	AD	1	1	1	3
8	CD	BD	BC	0	2	2	2

No matter what the players' abilities, there is always a three-time winner or a three-time loser.

23. Random Defective Components on Cards

16.7%. The probability of at least one defective component on a card is one minus the probability of none:

$$1 - (1 - p)^M = 1 - (1 - .03)^6 = .167, \text{ or } 16.7\%$$

This value is quite a bit higher that the 3% for components. With random defective components, the percent defective components must be very low to have a respectably low percent defective cards.

24. Clustering of Defective Components on Cards

3%. For N cards, the total number of components is MN, of which MNp are defective. If we herd M of these onto one card, M onto another, etc., then the number of cards affected is MNp/M, or Np. The fraction defective cards is Np/N, or p, the

same as the fraction defective components. The clustering drama-
tically reduces the fraction defective cards. Here $p = .03$, or 3%.

25. One Defective Component per Card

18%. Again, the total number of components is MN, of which
MNp are defective. Each defective component causes one defective card, so there are MNp defective cards too. The fraction
defective cards is MNp/N, or Mp. When $Mp > 1$, it means that
$MNp > N$, or the defective components outnumber the total
cards. There are more than enough defective components to go
around, so every card gets at least one. All the cards are defective.
Here Mp is $6(.03) = .18$, or 18%.

Problems 23–25 show that the fraction defective cards depends on how the defective components are distributed across
the cards. However, for p fraction defective components and M
components per card, bounds for the fraction defective cards are
p and Mp, with $1 - (1 - p)^M$ for a random distribution. In
general, this applies to components in any *system*, such as the
card here.

26. Shrimp Chow Mein

.040. Shrimp per serving, like defects per unit, should follow a
Poisson distribution. See Duncan (5): pp. 461–462. From his Table
E, we find that when the expected number is 5, the probability
of 1 or less is .040. Since this is less than the usual critical value
of .05 for a significant difference, we conclude that it's corporate
cost cutting, not chance.

27. Outgoing Percent Defective After 100% Screen

.11%. Let "bads" be units with the reliability defect and "goods"
be units without the reliability defect. This is really an inspection
error problem with

α = Fraction of true goods called bad at the screen = .05

β = Fraction of true bads called good at the screen = .05

See my "Sampling Plan Adjustment for Inspection Error and
Skip-Lot Plan," *Journal of Quality Technology*, July 1982, pp.

105–116. For N tested units, let p' be the true initial fraction defective for the reliability defect, and p be the expected value of the observed fraction defective at the screen. Observed defectives are either true bads correctly called bad, or true goods incorrectly called bad, so the relation between p and p' is:

$$p = \frac{p'(1 - \beta)N + (1 - p')\alpha N}{N} = \alpha + (1 - \alpha - \beta)p'$$

$$p' = \frac{p - \alpha}{1 - \alpha - \beta}$$

Here $\alpha = \beta = .05$ and $p' = .02$, so $p = .0680$. After the screen, the only reliability defectives shipped are units truly defective *and* missed at the screen, or $p'\beta N$. The N is reduced by pN at the screen, so the average outgoing quality (AOQ) as a percent defective is

$$\text{AOQ} = 100\frac{p'\beta N}{N - pN} = 100\frac{p'\beta}{1 - p} = .11\%$$

28. Yield for 100% Screen

93.20%. The yield (Y) is the percent of the starting units that pass the screen:

$$Y = 100\frac{N - pN}{N} = 100(1 - p) = 93.20\%$$

29. Quality of Scrap for 100% Screen

28%. A table of probabilities for the N tested units is helpful to do this problem and also to see the whole picture:

	True		
Called	Good	Bad	Total
Good	$(1 - p')(1 - \alpha)$	$p'\beta$	$1 - p$
Bad	$(1 - p')\alpha$	$p'(1 - \beta)$	p
Total	$1 - p'$	p'	1

This tells us immediately that the conditional probability that a unit is truly bad, given that it was called bad at the screen, is

$$\frac{p'(1 - \beta)N}{pN} = \frac{p'(1 - \beta)}{p} = .2794, \text{ or } 28\%$$

Only 28% of the units thrown out are actually bad, but the good units discarded may be considered the price of lowering the percent defective from 2% to .11%.

30. Even Number of Occurrences

$.5 + .5(q - p)^n$. The binomial theorem gernerates the probabilities for the binomial distribution:

$$(q + p)^n = \sum_{x=0}^{n} C(n,x)p^x q^{n-x}$$
$$= P(0) + P(1) + P(2) + P(3) + \ldots,$$

where $C(n,x) = \binom{n}{x} = n!/[x!(n - x)!]$ is the number of combinations of n things taken x at a time. It's also true that

$$(q - p)^n = \sum_{x=0}^{n} (-1)^x C(n,x)p^x q^{n-x}$$
$$= P(0) - P(1) + P(2) - P(3) + \ldots$$

It follows that

$$(q + p)^n + (q - p)^n = 2[P(0) + P(2) + P(4) + \ldots]$$
$$.5 + .5(q - p)^n = P(0) + P(2) + P(4) + \ldots$$

31. Bit Lines and Vias

$.000016$. Assume the bad vias are spread randomly over the chips, so that $x = $ bad vias on a bit line follows a binomial distribution with $n = 256$ vias and $p = $ the unknown probability that a via is bad. The probability of x greater than or equal to 2 can be estimated from the sample:

$$P(x \geq 2) = \frac{\text{Bad lines}}{\text{Total lines}} = \frac{3}{331(1152)} = .0000079$$

Now use systematic trial and error with the binomial distribution until you find the p such that $P(x \geq 2)$ is .0000079. This p is the required fraction bad vias. The trial and error can be done on APL. Execute)COPY 42 STAT6 BINOCDF, and then the probability of x or less is given by p BINOCDF n x. We want $P(x \geq 2)$ to be .0000079, or equivalently $P(x \leq 1)$ to be $1 -$ .0000079 = .9999921. Hence, we use p BINOCDF 256 1 until the value returned is as near as possible to .9999921. To 6 decimal places, the best value is p = .000016, or 16 parts per million.

32. Clusters

511. In general, think of taking some number of objects out of the initial group of n objects, so the two subgroups are those taken and those not. Each of the n objects can be taken or not, so there are 2^n total ways to make the selection. Excluding none or all selected gives $2^n - 2$, but this counts each division twice. For example, if the objects are A, B, and C, then A selected is the same as B and C selected. The ways to form two subgroups then is $(2^n - 2)/2$, or $2^{n-1} - 1$. For $n = 10$, this gives 511.

33. NCAA Basketall Tournament

51. Each game eliminated one team, and 51 teams had to be eliminated.

34. Password

1,212,640. There are 26 letters and 10 digits, so there are 36 ways to fill the first position, 36 ways to fill the second position, etc. This suggests 36^4 different permutations, but this includes the all-letters and all-digits cases, which are not allowed. Subtracting these out gives the answer: $36^4 - 26^4 - 10^4 = 1{,}212{,}640$.

35. Trimming a Tree

255. Consider the general tree with maximum number of n. If we start at the 1 and go left or right *only* to a 2, there are 2 choices for this. After that, there are 2 ways to go to a 3, then 2 ways to a 4, etc. Multiplying the choices together gives 2^{n-1} paths. Now

go from the 1 *up* to the 2, then left or right *only* to a 3, and no restrictions after that. This gives 2^{n-2} paths. Continuing in this manner, we see the total number of paths is

$$2^{n-1} + 2^{n-2} + 2^{n-3} + \ldots + 2 + 1$$

This is the sum of the first n terms of a geometric progression with ratio 2, which reduces to $[1 - 2^n]/[1 - 2]$, or $2^n - 1$. For $n = 8$, we have 255.

36. Combinations of Chemicals

16. For each of the four chemicals, there are two choices: present or not present. Hence, the total number of combinations is $2(2)(2)(2)$, or $2^4 = 16$.

37. Combinations of Chemicals (*continued*)

2. The number of combinations of n things taken r at a time is $C(n,r) = n!/[r!(n - r)!]$ An interesting display of $C(n,r)$ is given by Pascal's triangle. Note that each row gives $C(n,r)$ for $r = 0$, $1, 2, \ldots , n$. Also, each value is the sum of the two above it. If n is even, the maximum $C(n,r)$ is when $r = n/2$. For odd n, $C(n,r)$ is the same maximum value for the integers above and below $n/2$. Here with $n = 4$, the maximum is $C(4,2) = 6$.

n									
0					1				
1				1		1			
2			1		2		1		
3		1		3		3		1	
4	1		4		6		4		1
5	1	5		10		10		5	1
6	1	6	15		20		15	6	1
7	1	7	21	35		35	21	7	1

38. Testing All Combinations of Chemicals

6. This is the maximum value of $C(4,r)$ and occurs for $r = 2$. The one-chemical additions can be tested on the way to the six two-

chemical combinations, and these six samples are enough to test the combinations with r greater than two. In order to run the experiment, you have to write out the combinations, and one way to do this would be:

			r = Chemicals added		
Sample	0	1	2	3	4
1		A	B	C	D
2		B	C	D	
3		C	A	D	
4		D	C		
5		A	D	B	
6		B	D		
Combinations	1	4	6	4	1

39. Seating Diplomats

120. Suppose the positions were distinguishable, as they would be if the diplomats were seated in a row at a head table. Then we would say there are 6 ways to fill the first position, 5 ways to fill the second, etc., for a total of 6! = 720 ways. But here the positions are indistinguishable. To make them distinguishable, consider one diplomat and the relation of the others to him. There are 5 ways to fill the seat on his left, 4 ways to fill the next seat to the left, etc., for a total of 5! = 120 ways, which is the answer. For each of these 120 ways, everyone can move 0, 1, . . . , 5 seats to the left for 6 arrangements that are all the same. If they were different, the answer would be 6(120) = 6! = 720.

40. Two Packs of Life Savers

705,432. There are 11 + 11 = 22 places for the order of eating. You can pick the Butter Rum places in $C(22,11)$ ways, where $C(n,r) = n!/[r!(n - r)!]$ is the number of combinations of n things taken r at a time. The Wild Cherry must go in the remaining places. The answer then is $C(22,11)$, which works out to be 705,432.

41. Life Savers Puzzle

2×10^{12}, or 2 trillion. According to a fundamental principle, if one thing can be done in m ways, and then another thing can be done in n ways, then the two things can be done together mn ways. Here there are 12 ways to pick the first piece and 2 ways to place it (side A up or side B up). Then there are 11 ways to pick the next piece and 2 ways to place it. This continues through all 12 pieces, which gives

$$(12 \cdot 2)(11 \cdot 2) \ldots (1 \cdot 2) = 2 \times 10^{12}$$

42. Driving to Work

12,870. The North and East segments can be driven in any order, for instance NNENEE . . . , as long as there are eight Ns and eight Es. You can pick the places for the Ns in C(16,8) ways, where the number of combinations of n things taken r at a time is

$$C(n,r) = \frac{n(n-1)(n-2) \ldots \text{to } r \text{ terms}}{r!}$$

The Es must go in the remaining places, so the answer is C(16,8) = 12,870. If you drive to work and back 230 times a year, the one-way drives per year would be 2(230) = 460, so you could cover all the routes in one direction or the other in 12,870/460 = 28 years.

43. Walking to Work

6.0×10^8. This can be handled just like Problem 42, provided you have the insight to see that it really doesn't matter that the length of a street or avenue crossing is different from the length of a block side. You must have eight crossings and eight block sides to the north, and the same to the east. Hence, there are sixteen Ns and sixteen Es, so the answer is C(32,16), which is 6.0×10^8. You will probably retire before getting a chance to try all the routes.

44. Matching Socks

$51/91 = .56$. The number of combinations of n things taken r at a time is

$$C(n,r) = \frac{n(n - 1)(n - 2) \ldots \text{to } r \text{ terms}}{r!}$$

The total equally likely ways to pick the 2 socks is $C(14,2)$. Of these, the favorable ways from the 5 pairs of one kind is $C(10,2)$, and the favorable ways from the 2 pairs of the other kind is $C(4,2)$. The required probability is the ratio of favorable ways to total ways, which reduces to $51/91 = .56$.

45. Matching Socks (*continued*)

$3/143 = .021$. Call the 10 socks of one kind As, and the 4 others Bs. Consider the order that the socks come out of the drawer. There are $C(14,4)$ ways to pick the B positions. Then the As must go in the other 10 positions, so the total equally likely orders is $C(14,4)$. The favorable orders have Bs in 2 of the following 7 pairs of positions: (1,2), (3,4), (5,6), (7,8) (9,10), (11,12), and (13,14). The ways to pick the 2 pairs of positions is $C(7,2)$, and then the As must go in the other 10 positions. The required probability is the ratio of favorable ways to total ways, which reduces to $3/143 = .021$.

46. Choosing Desserts

2^n. The binomial theorem states that

$$(a + b)^n = \sum_{x=0}^{n} C(n,x)b^x a^{n-x}$$

If we let $a = b = 1$, we have

$$2^n = \sum_{x=0}^{n} C(n,x)$$

Hence, 2^n is the total number of ways to pick 0, 1, 2, . . . items from n different items.

47. 5-Button Combination Lock

1/420. The required answer is one divided by the total number of possible combinations. Since order is important, these are really permutations, not combinations. It should be called a permutation lock! For 3 entries, there are $\binom{5}{2} = 10$ ways to pick the double, then $\binom{3}{2} = 3$ ways to pick the single digits, and $3! = 6$ ways to order the double and the 2 digits. Remember that if one thing can be done in m ways, and then another thing can be done in n ways, then the two things can be done together in mn ways. This means the number of three-entry "combinations" is $10(3)(6) = 180$. Similar reasoning gives the number of four-entry combinations as $\binom{5}{2}\binom{3}{3}4! = 10(1)(24) = 240$. The total number of combinations is 180 plus 240, or 420, so 1/420 is the answer.

48. Fish on Venus

5. This is a variation on the classical balls-in-cells problem, given in Feller (6). In effect, the astronaut places r fish at random in the three unknown sex cells A, B, and C. The probability of an ordered selection like BAAC . . . is $(1/3)^r$. To get the success probability, we multiply this by the number of selections with at least one fish of each sex. Suppose $r = 4$, and consider the *unordered* cell counts 1 1 2—that is, the first number is not necessarily the A count. The numbers are written from low to high for convenience. Let O_1 be the ways to order the cells. For example, one order is A $= 1$, B $= 2$, and C $= 1$. Then let O_2 be the ways to get a particular set of ordered cell counts. For A $= 1$, B $= 2$, and C $= 1$, one way would be BACB. To evaluate O_1 and O_2, we need to know that the number of ways to order n things, of which r_1 are alike of one kind, r_2 alike of another, etc., is $n!/(r_1!r_2! \ldots)$. Finally, the probability of the unordered cell counts 1 1 2 is $O_1O_2(1/3)^4$. For $r = 4$, this is the only favorable set. In general, there may be more than one favorable set, so the success probability is

$$(\Sigma O_1O_2)(1/3)^r = \Sigma O_1O_2(1/3)^r,$$

where the summation is understood to be over all favorable sets. The table shows the possibilities:

r	Unordered cell counts	O_1	O_2	$O_1 O_2 (1/3)^r$
3	1 1 1	1	$3! = 6$	$\dfrac{6}{27} = .22$
4	1 1 2	$\dfrac{3!}{2!1!} = 3$	$\dfrac{4!}{1!1!2!} = 12$	$\dfrac{36}{81} = .44$
5	1 2 2	$\dfrac{3!}{1!2!} = 3$	$\dfrac{5!}{1!2!2!} = 30$	$\dfrac{90}{243}$
	1 1 3	$\dfrac{3!}{2!1!} = 3$	$\dfrac{5!}{1!1!3!} = 20$	$\dfrac{60}{243}$
				$\dfrac{150}{243} = .62$

Since .62 is over .5, $r = 5$ is the required answer. Feller shows that for r balls in n cells, the probability that exactly m cells remain empty is

$$p_m(r,n) = \binom{n}{m} \sum_{v=0}^{n-m} (-1)^v \binom{n-m}{v} \left(1 - \frac{m+v}{n}\right)^r,$$

where $\binom{n}{m} = n!/[m!(n-m)!]$ is the number of combinations of n things taken m at a time. With $r = 5$, $n = 3$, and $m = 0$, this gives the .62. For a less frivolous application, consider a random sample of r units from the thoroughly mixed output of n tools, and we want 90% confidence that each tool has at least one unit in the sample.

49. Blackjack

21. The first card can be an ace and the second a 10, or vice versa. Summing the two mutually exclusive probabilities gives the probability of getting a 21:

$$\frac{16}{208}\left(\frac{64}{207}\right) + \frac{64}{208}\left(\frac{16}{207}\right) = .0476$$

Setting .0476 = $1/n$ gives n = 21, so 21 happens once every 21 hands.

50. Winning at Craps

244/495 = .493. Think of one die as being red and the other black. Then with R for red and B for black, the (R,B) ways to get a 7 are (1,6), (6,1), (2,5), (5,2), (3,4), and (4,3). These six mutually exclusive ways all have the same (1/6)(1/6) = 1/36 probability, so the probability of a 7 is (6)(1/36) = 6/36. Other possibilities are

Dice total	2	3	4	5	6	7	8	9	10	11	12
36 × Prob.	1	2	3	4	5	6	5	4	3	2	1

This table tells us that the chance of winning on the first roll is (6 + 2)/36, or 8/36. Also, the chance of a 4 on the first roll is 3/36. Then there are 3 + 6 = 9 equally likely ways for the game to end, and the chance of "making your point" is 3/9. The chance of this joint occurrence, 4 on the first roll and then 4 before 7, is (3/36)(3/9). Summing the chances for all the point possibilities gives

$$\frac{3}{36}\left(\frac{3}{9}\right) + \frac{4}{36}\left(\frac{4}{10}\right) + \frac{5}{36}\left(\frac{5}{11}\right) + \frac{5}{36}\left(\frac{5}{11}\right) + \frac{4}{36}\left(\frac{4}{10}\right) + \frac{3}{36}\left(\frac{3}{9}\right)$$

To this, we add the first-roll chance of 8/36. Careful canceling gives 244/495 = .493.

51. 6 Before 7

5/11. The ways to get a 6 are (1,5), (5,1), (2,4), (4,2) and (3,3), each with probability (1/6)(1/6) = 1/36. The overall probability of a 6 is the sum of the probabilities for these mutually exclusive events, or 5/36. Similarly, the probability of a 7 is 6/36. In n rolls, with n indefinitely large, (5/36)n will be a 6, and (11/36)n will be a 6 or 7. These are the only totals that count, so the probability that a 6 is the first roll that counts is

$$\frac{\left(\dfrac{5}{36}\right) n}{\left(\dfrac{11}{36}\right) n} = \frac{5}{11}$$

52. Like Roulette?

$30.94, every time! No matter what the order of the cards, the house's money is multiplied by $(1 - .20)$ a total of 18 times and by $(1 + .20)$ a total of 20 times. Hence, at the end the house has $(.8)^{18}(1.2)^{20}(\$100)$, which is $69.06. The missing $30.94 must be in your pocket. As you go through the deck, note that if the house is winning, the remaining cards must be red-rich, and you bet *more*. If the house is losing, the remaining cards must be red-poor, and you bet *less*. This is the key. Your bet varies directly with your probability of winning. Unfortunately, Vegas does not offer this game. The roulette wheel is always red-poor due to the zero and double-zero.

53. Winning a Kewpie Doll

$4.75. On the average, you win once in 20 times, so the prior 19 losses cost you 19 ($.25), or $4.75.

54. Pregame Coin Flips

.007. We have a binomial distribution here with $n = 15,000$ teams and $p = (1/2)^{21}$. Since p is so small, the Poisson distribution will be an excellent approximation, with the probability of zero occurrences equal to e^{-np}, or .993. The probability of at least one occurrence is one minus this, or .007. Hence, the Bloomington team's losing streak *is* rare, but not as rare as $(1/2)^{21} = 5 \times 10^{-7}$ suggests.

55. Beer Tasting

1/24. Suppose you guess two right. Then you break even, because $2 - $2 = $0. Hence, you must get three right to make some money. But if three are right, the fourth one is too, so you must guess all four right! There are 4! = 24 different orders for the four beers, only one of which is right, so the answer is 1/24.

56. Beer Tasting (*continued*)

−$2. Let A, B, C, and D be the four brands ordered from left to right in front of you. There are 4! = 24 equally likely ways to guess:

Number right	x = Winnings ($)	Ways	$P(x)$
0	−4	BADC BCDA BDAC CADB CDAB CDBA DABC DCAB DCBA	$\frac{9}{24}$
1	−2	ACDB ADBC CBDA DBAC BDCA DACB BCAD CABD	$\frac{8}{24}$
2	0	ABDC ADCB ACBD DBCA CBAD BACD	$\frac{6}{24}$
3	—	—	$\frac{0}{24}$
4	4	ABCD	$\frac{1}{24}$

The average winnings per game is $\Sigma x P(x)$ = − $2.

57. Three Cards

No. Originally, there was a 2 out of 3 chance of picking a card with the same letter on each side. The fact that he picked a card with a Y doesn't change the probabilities because he would have suggested the same bet if an X had appeared. Hence, the real odds are still 2 to 1 for a match, so he should wager $2 to your $1 for a fair bet. It might be thought that the chance of a Y under is 1/2, since the card has to be one of two cards: (X,Y) or (Y,Y). However, given a Y up, these two cards are not equally likely.

The equally likely possibilities are:

Side	Card 1		2
Up	Y_1	Y_2	Y
Under	Y_2	Y_1	X

Again, the chance of a match is 2/3.

58. Choke?

.04. Treat each shot as an independent trial with the probability of a miss equal to $1 - .80$, or .20. Then the probability of two consecutive misses is $(.20)^2 = .04$. Since this is less than .05, the event is significant at the .05 level. In other words, he choked.

59. Another Choke?

Yes. We can use Fisher's Exact Test on these two proportions. On APL, execute)COPY 42 STAT3 FISHERΔEXACT PROMPT, and then 9 15 FISHERΔEXACT 3 16. This returns a one-tail probability of only .023, which ordinarily would be significant. However, it is well known that Celtics don't choke, so this must be just due to chance. For more on Fisher's Exact Test, see Problems 126, 174–177, 184, and 188.

60. Repeated Number in Lottery

.79. On any day there are 1000 equally likely numbers, only one of which is yesterday's, so the chance of a repeat is 1/1000. With 52 Sundays in a year, the number of drawings in a five-year period is 5(365−52), or 1565. The probability of at least one repeat is one minus the probability of no repeats, or

$$1 - (1 - .001)^{1565} = .79$$

The average waiting time in years for a repeat is drawings/(drawings/year) = 1000/(365 − 52), or 3.2 years.

61. House Edge for Lottery

50%. The probability that a player wins is 1/1000, so for every 1000 bets of $1 each, there is only one winner, on the average. This means the state takes in $1000 and pays out $500, for a house edge of $(1000 - 500)/1000 = .50$, or 50%. More formally, let x be the house gain on a single $1 bet, and $P(x)$ be the probability of x. Then the house edge is

$$100 \sum xP(x) = 100\left[(1)\left(\frac{999}{1000}\right) + (-499)\left(\frac{1}{1000}\right)\right] = 50\%$$

By contrast, the house edge for the casino game of roulette is only 5%.

62. A Statistician's Urn

White. With B for black and W for white, possibilities on a draw are:

| | Change in count | |
Draw	White	Black
BB	0	-1
WW	-2	$+1$
BW or WB	0	-1

Note that the white count either stays the same or is reduced by two. Since the white count is originally odd, it *stays* odd and eventually decreases to one. If there are any black balls left at that point, then following draws must be BB, BW, or WB. These all decrease the black count by one until only the white ball is left.

63. Another Statistician's Urn

1/2. The probability of winning by n trials could be found by adding the probabilities of winning on the first trial, the second trial, etc., but we can avoid this infinite series by taking one minus the probability y of *not* winning at any of the *n* trials:

$$y = \left[1 - \left(\frac{1}{2}\right)^2\right]\left[1 - \left(\frac{1}{3}\right)^2\right] \cdots \left[1 - \left(\frac{1}{n+1}\right)^2\right]$$

$$= \left[\frac{2^2 - 1}{2^2}\right]\left[\frac{3^2 - 1}{3^2}\right] \cdots \left[\frac{(n+1)^2 - 1}{(n+1)^2}\right]$$

$$= \left[\frac{(2-1)(2+1)}{2^2}\right]\left[\frac{(3-1)(3+1)}{3^2}\right] \cdots \left[\frac{n(n+2)}{(n+1)^2}\right]$$

For example, suppose that $n = 6$. In what follows, lines connect terms that completely or partially cancel, and the circled values are left:

$$y = \frac{1 \cdot 3}{②^2} \diagup \frac{2 \cdot 4}{3^2} \diagup \frac{3 \cdot 5}{4^2} \diagup \frac{4 \cdot 6}{5^2} \diagup \frac{5 \cdot 7}{6^2} \diagup \frac{6 \cdot ⑧}{⑦^2}$$

$$= \frac{8}{2(7)} = \frac{n+2}{2(n+1)}$$

Consultation with Leon Lareau, IBM Burlington, led to this alternative to the canceling:

$$y = \prod_{i=1}^{n} \frac{i(i+2)}{(i+1)^2}$$

$$= \left(\prod_{i=1}^{n} i\right) \frac{\prod_{i=1}^{n}(i+2)}{\prod_{i=1}^{n}(i+1)^2}$$

$$= n! \frac{\dfrac{(n+2)!}{2}}{[(n+1)!]^2}$$

$$= \frac{n!(n+2)(n+1)!}{2(n+1)(n!)(n+1)!}$$

$$= \frac{n+2}{2(n+1)}$$

The probability of winning at any of the n trials is $1 - y$, or $n/(2n + 2) = 1/[2 + (2/n)]$. Taking the limit as n goes to infinity gives the answer of 1/2. It might be thought that since the player can always play one more trial, his probability of eventually winning must be one. However, this is not true. At the nth trial, the probability of winning is $[1/(n + 1)]^2$, which goes to 0 as n goes to infinity. Just because you try something for an indefinitely long time doesn't mean eventual success. Try flipping a coin onto a table top so that it ends up on its edge.

64. Yet Another Statistician's Urn

1. The probability of winning by n trials could be found by adding the probabilities of winning on the first trial, second trial, etc., but we can avoid this infinite series by taking one minus the probability y of *not* winning at any of the n trials:

$$y = \left(1 - \frac{1}{2}\right)\left(1 - \frac{1}{3}\right) \cdots \left(1 - \frac{1}{(n + 1)}\right)$$

$$= \frac{1}{2}\left(\frac{2}{3}\right) \cdots \left(\frac{n}{n + 1}\right)$$

$$= \frac{n!}{(n + 1)!}$$

$$= \frac{1}{n + 1}$$

Taking the limit as n goes to infinity gives $y = 0$, so the probability of winning is one.

65. Cutting for the Deal

1/221. The chance that the first player cuts an ace is 4/52. This leaves 3 aces in 51 cards for the second player, so his probability is 3/51. Multiply the probabilities to get the answer:

$$\frac{4}{52}\left(\frac{3}{51}\right) = \frac{1}{221}$$

66. Hexogram

45. First consider the pieces with the three sections all the same color. There must be 5 of these, one for each color. Now consider the pieces with two sections of one color and the third another. There are 5 colors for the matching sections, and then 4 colors for the third section, for a total of 5(4) = *20* pieces. Finally, consider the pieces with the three sections all different. The ways to pick the three colors is the number of combinations of 5 things taken 3 at a time, or 5(4)(3)/3(2) = 10. Once we have the three colors, it's like putting diplomats around a circular conference table (Problem 39). Consider one color and the relation of the others to it. There are two choices for the color to its left, and then only one choice for the other one. This gives only two ways to order the three colors, so that the number of pieces with the sections all different is 10(2) = *20*. The total number of pieces then is 5 + 20 + 20 = 45.

67. Yarborough in Bridge

1827 to 1. If the probability of an event is p, then the odds against the event is the ratio of unfavorable ways to favorable ways, or $(1 - p)/p$. Find p first. The number of combinations of n things taken r at a time is $C(n,r) = n!/[r!(n - r)!]$. There are 32 cards 2 through 9, so the probability of a Yarborough is

$$p = \frac{C(32,13)}{C(52,13)}$$

$$= \frac{32(31)(30) \ldots (20)}{52(51)(50) \ldots (40)}$$

Careful canceling gives

$$p = \frac{31(3)(2)(29)}{17(7)(47)(43)(41)}$$

Since all the factors are prime numbers, no more canceling is possible, and $p = 5394/9,860,459$. The required odds are

$$\frac{1 - p}{p} = \frac{9,860,459 - 5394}{5394} = \frac{1827}{1}$$

My guess is that the Earl of Yarborough knew these fair odds and shrewdly offered the bet with lower odds.

68. Perfect Cribbage Hand

1/216,580. First, consider the probability that your six-card hand gives you the opportunity for a perfect hand. The total number of six-card hands is $C(52,6)$, where $C(n,r) = n!/[r!(n - r)!]$ is the number of combinations of n things taken r at a time. This is more easily evaluated as

$$C(n,r) = \frac{n(n - 1)(n - 2) \ldots \text{to } r \text{ terms}}{r!}$$

For the favorable hands, remember a fundamental principle that if one thing can be done in m ways, and then another thing can be done in n ways, then the two things can be done together mn ways. There are $C(4,3)$ ways to get the 5s. Then the jack must be the same suit as the missing 5. The other two cards can be anything except this jack and the four 5s, so there are $C(52 - 5,2)$ ways to get the two. The probability of a favorable six-card hand then is

$$\frac{[C(4,3)][1][C(47,2)]}{C(52,6)}$$

This must be multiplied by the probability that the cut card is the fourth 5, or $1/(52 - 6)$. Careful canceling gives 1/216,580.

69. Probability of Player's Ruin

.94. See Feller (6). In general, suppose a player with initial capital z plays until he either loses it all or increases it to an amount a. The z and a are in *units of bet size*. If the player's chance of winning a single bet is p, and $q = 1 - p$, then Feller shows that his ruin probability is

$$
q_z = \frac{\left(\dfrac{q}{p}\right)^a - \left(\dfrac{q}{p}\right)^z}{\left(\dfrac{q}{p}\right)^a - 1}
$$

If $p = q = 1/2$, then this gives $q_z = 0/0$, and Feller shows that $q_z = 1 - (z/a)$ in this special case, which unfortunately won't happen in a casino. Here the bet size is $1, so $z = 900$ and $a = 1000$. With $p = 244/495$ or .493 from Problem 50, we find $q_z = .94$.

70. Probability of Player's Ruin (*continued*)

.11. If the bet size is $100, then $z = 900/100 = 9$ and $a = 1000/100 = 10$. The q_z formula now gives .11, a dramatic reduction from the .94 for the bet size of $1. Contrary to popular opinion, it *is* possible to go up against the casino and have a high chance of winning.

71. Celtics Home and Away Records

.00001. The split between the two win/lose cells is 20 to 1. If chance is the only factor, then we expect an equal split, like flipping coins. This suggests a binomial distribution model with $n =$ trials $= 21$ and $p =$ probability of occurrence $= .5$, which gives $P(1$ or less$) = .00001$. On APL, this can be found with)LOAD 42 STAT6, and then .5 BINOCDF 21 1 returns the $P(1$ or less). In Siegel (15), this approach to the problem is discussed under the McNemar Test for the significance of changes. It might be thought Fisher's Exact Test could be used for two proportions on the home and away records, but the samples are not independent, since the opponents are the same home and away. Another incorrect approach would be to reason that you expect half the losses at home and half away, so the binomial distribution with $n = 23$ losses and $p = .5$ gives $P(2$ or less$) = .00003$. This is close to the .00001 answer, but if we do a similar analysis on the wins, with $n = 59$ and $p = .5$, we get a different answer: $P(20$ or less$) = .00917$. Something must be wrong, and what's wrong is treating the wins or losses as independent trials with $p = .5$ for each trial. It's impossible, for example, to get more than 41 wins at home because there are only 41 games played at home.

72. Celtics Home and Away Stats

.008, .134, .027, .017, .017, .412, .027. All the statistics are significant except Free-Throw % and Technical Fouls. The conclusion is that at home the Celtics play better *and* the refs favor them. The edges are small, but they add up to a big advantage. The one-tail probabilities are found by using the binomial distribution with n = trials = number of pluses and minuses, and p = probability of occurrence = .5. An occurrence is a sign in the anticipated direction, which is plus for everything except the fouls. As an example, for Rebounds we have $n = 39$ and $P(26$ or more$) = .027$, which could be evaluated as $1 - P(25$ or less$)$ or $P(13$ or less$)$.

73. Two-Time Loser

1/6. The long solution is to reason that Ann can lose both times, or Bill can lose both times, or . . . The required answer is the sum of the probabilities for these six mutually exclusive events:

$$\left(\frac{1}{6}\right)\left(\frac{1}{6}\right) + \left(\frac{1}{6}\right)\left(\frac{1}{6}\right) + \ldots = 6\left(\frac{1}{6}\right) = \frac{1}{6}$$

The short solution is to realize that the morning loser has a 1/6 chance of losing in the afternoon, and that's the answer.

74. Russian Roulette

The possibilities and their probabilities can be presented in a table, with "random" for the spinning of the cylinder:

	Placement	
Pulls	Random	Consecutive
Random	$\left(\dfrac{4}{6}\right)\left(\dfrac{4}{6}\right)$ = .44	$\left(\dfrac{4}{6}\right)\left(\dfrac{4}{6}\right)$ = .44
Consecutive	$\left(\dfrac{4}{6}\right)\left(\dfrac{3}{5}\right)$ = .40	$\left(\dfrac{4}{6}\right)\left(\dfrac{3}{4}\right)$ = .50

The cylinder spin before the first pull gives a random chamber. Hence, the survival probability for the first pull is always (favorable chambers)/(total chambers), or 4/6. This is multiplied by the survival probability for the second pull, which again is 4/6 if the second pull is random, regardless of placement method. If the pulls are consecutive, then the placement method makes a difference. Random placement means 3 favorable chambers left out of the total of 5 left, so the chance that the next one is empty is 3/5. For consecutive placement, call the empty chambers 1 through 4, and the full ones 5 and 6. Let the possible pulls be (1,2), (2,3), (3,4), (4,5), (5,6), and (6,1). Then if the first pull is for chambers 1, 2, or 3, the second pull is for an empty chamber. Given that the first pull was for an empty chamber, the conditional probability that it was 1, 2, or 3 is (favorable empty chambers)/(total empty chambers), or 3/4. Alternatively, the chance of surviving *both* pulls is the chance that the first pull is 1, 2 or 3, which leads directly to the 3/6 = .50 probability.

75. Momentum

75%. The Game 5 winner wins the Series if it wins the sixth game, or if it loses the sixth game but wins the seventh game. With L for lose and W for win, the probability of these two mutually exclusive events is

$$P(W) + P(LW) = .5 + .5(.5) = .75, \text{ or } 75\%$$

The observed 71% agrees quite well with this. Support for the momentum theory would be an observed percent significantly greater than 75%. Can it be that the Associated Press compared the observed 71% against 50%, forgetting that the one-game edge changes the probability of winning the Series? By the way, the Twins won the Series.

76. Sudden Death

2/3. In general, let p be the probability that a team scores on a possession. Then the probability that it doesn't score is $q = 1 - p$. The receiving team wins if it scores on its first possession, or both teams fail to score on their first possession, and the receiving team scores on its next possession, etc. The addition and multiplication probability rules give the probability of this:

$$p + q^2p + q^4p + q^6p + \ldots$$
$$p[1 + q^2 + (q^2)^2 + (q^2)^3 + \ldots]$$

The expression in the brackets is a geometric series of the general form

$$1 + x + x^2 + x^3 + \ldots$$

If $x < 1$, the series reduces to $1/(1 - x)$. Here $x = q^2$, so the probability that the receiving team wins is $p/(1 - q^2)$. Substituting $p = q = 1/2$ gives 2/3.

77. Rolling a Die

3. In general, if the results of a trial are equally likely, and an occurrence is one of the favorable results, then the probability of an occurrence is $p = $ (favorable results)/(total results). Here a trial is a roll, the 6 equally likely results are the numbers 1 through 6, and the 2 favorable results are 3 and 6, so $p = 2/6 = 1/3$. It's also true that $p = $ occurrences/trials for an indefinitely large number of trials. The average number of trials per occurrence, which is the average number of trials to the first occurrence, is trials/occurrences $= 1/p$. Hence, the required average is $1/p = 1/(1/3) = 3$.

78. Rolling a Die (*continued*)

14.7. After $i - 1$ numbers have been achieved, let x_i be the number of rolls to get the ith new number, with $i = 1, 2, \ldots, 6$. If y is the number of rolls to get all six numbers, then it must be true that

$$y = x_1 + x_2 + \ldots + x_6$$

Since the average of a sum is the sum of the averages, the average value of y is

$$\mu_y = \mu_1 + \mu_2 + \ldots + \mu_6,$$

where μ_i is the average value of x_i. In Problem 77, we saw that $\mu_5 = 3$. In general, if p_i is the probability of rolling the ith new number, after $i - 1$ numbers have been achieved, then

$$p_i = \frac{6 - (i - 1)}{6},$$

and $\mu_i = 1/p_i$. This leads to

$$\mu_y = \frac{6}{6} + \frac{6}{5} + \frac{6}{4} + \frac{6}{3} + \frac{6}{2} + \frac{6}{1} = \frac{882}{60} = \frac{147}{10} = 14.7$$

79. Six Different Numbers in Six Rolls

5/324. Any number will do for the first number, but after that, we must have no repeats. This means there are 5 favorable numbers for the second roll, 4 favorable numbers for the third roll, etc. Multiplying the probabilities gives the probability of the joint occurrence of anything on the first roll, *and* no repeat on the second roll, *and* no repeat on the third roll, etc.:

$$\frac{6}{6} \left(\frac{5}{6}\right)\left(\frac{4}{6}\right)\left(\frac{3}{6}\right)\left(\frac{2}{6}\right)\left(\frac{1}{6}\right) = \frac{5}{324}$$

The average number of tries to achieve this is $1/(5/324) = 324/5 = 65$.

80. A Rich Uncle

It makes no difference what you do. One way to see this is to realize that whether you trade or not, you have a random pick from the two envelopes, so there can't be any difference. Formally, suppose one envelope has D dollars and the other $2D$ dollars. Let x be the gain by trading, and $P(x)$ be the probability of x. Then the expected value of x, which is the long-run average value of x, is

$$E(x) = \Sigma x P(x) = .5(2D - D) + .5(D - 2D) = 0$$

81. A Rich Aunt

Yes. Suppose the envelope you have contains D dollars, so the one you trade for has $D/2$ dollars or $2D$ dollars. If x is the gain by trading, and $P(x)$ is the probability of x, then the expected value of x is

$$E(x) = \Sigma x P(x) = .5\left(\frac{D}{2} - D\right) + .5(2D - D) = .25D$$

On the average, you gain 25% by trading, so do it! Some people mistakenly do this same analysis for Problem 80. Note there are three amounts of money involved here, but only two in Problem 80. They are different problems, although they may seem the same. In both problems, you may trade your envelope for one with either half as much or twice as much. However, in Problem 80, it's D vs. $2D$ *or* $2D$ vs. D, while in Problem 81, it's D vs. $D/2$ or $2D$.

82. Progression in Football Standings

Yes. Number the teams from 1 to 8 in an arbitrary manner, and consider the chance that the teams finish in that same order in the standings with 7–0, 6–1, . . . , 0–7 records. The chance that team 1 wins all its games is $(1/2)^7$. Team 2 has already lost to team 1, and the chance it wins its other 6 games is $(1/2)^6$. Team 3 has already lost to teams 1 and 2, and the chance it wins its other 5 games is $(1/2)^5$. Continue in this manner to team 7, which has already lost to teams 1 through 6, and the chance it wins its game with team 8 is 1/2. Team 8 has lost all its games, so there's no term for it. The chance that all this happens is the product of the probabilities. This is just one arbitrary order of the teams though, and each of the 8! orders has the same probability. The overall probability then is

$$\left(\frac{1}{2}\right)^7 \left(\frac{1}{2}\right)^6 \left(\frac{1}{2}\right)^5 \left(\frac{1}{2}\right)^4 \left(\frac{1}{2}\right)^3 \left(\frac{1}{2}\right)^2 \left(\frac{1}{2}\right)(8!)$$

Careful canceling gives 315/2,097,152 = .0001502 . . . Setting this equal to $1/n$ and solving for n gives n = 2,097,152/315 = 6658. *SI* must have calculated 1/.000150 = 6667. On the alleged rarity of the progression, *SI* says, "However, because in real life skill comes into play and some teams are always stronger than others, such perfect progressions actually occur far more frequently than that. In fact, the Big Eight's final standings were also in perfect progression in 1986."

83. Bingo Cover-All

.00010. Imagine that the 24 balls with the numbers on the card are red, and the others are white. We want the probability that we get all 24 red balls in a random sample of 55 balls from the total of 75 balls. This calls for the hypergeometric distribution with N = 75 total balls, k = 24 red balls, n = 55 balls drawn, and $x = k$ = 24 red balls in the sample. In general, the probability of exactly x red balls is given by

$$P(x) = \frac{\left(\begin{array}{c} k \\ x \end{array}\right)\left(\begin{array}{c} N - k \\ n - x \end{array}\right)}{\left(\begin{array}{c} N \\ n \end{array}\right)}$$

where the $\left(\begin{array}{c} k \\ x \end{array}\right)$ = $k!/[x!(k - x)!]$ is the number of combinations of k things taken x at a time. The x must be greater than or equal to the bigger of 0 and $n - (N - k)$, and also less than or equal to the smaller of n and k. Here $x = k$, so we have

$$P(k) = \frac{\binom{N - k}{n - k}}{\binom{N}{n}} = \frac{\binom{51}{31}}{\binom{75}{55}}$$

On APL, $X!K$ gives the number of combinations of X things *out of* K, so $(31!51) \div (55!75)$ gives the answer of $.0000965\ldots$, or $.00010$. Since APL operates from right to left and within parentheses first, we really don't need the second set of parentheses.

84. Bingo Cover-All (*continued*)

$24/75 = .32$. There are 75 equally likely numbers for the last one drawn, and you have 24 of them. Hence, the required probability is

$$\frac{\text{Favorable outcomes}}{\text{Total outcomes}} = \frac{24}{75} = .32$$

85. Random Sample

Get a deck of playing cards and take ace through 10 from two suits and ace through 5 from another. Renumber these cards from 1 to 25. For each random sample, first shuffle the cards well. Then thumb through the face-up deck and take out the first n cards that are less than or equal to N. These are the n wafer numbers for the random sample. Another method would be to work your way through a table of random numbers between 1 and 25. In APL, a table of 1000 such numbers could be generated by ?50 20ρ25.

86. Sampling Tool

Yes. This is a goodness-of-fit problem that can be handled with the chi-square test. See Duncan (5), pp. 634–637. Although the binomial distribution is a good approximation here since $n/N <$

.10, we will use the hypergeometric distribution, which takes into account the finite population size $N = 1000$ and population yellows, $k = 1000(.01) = 10$. On APL, execute)COPY 42 STAT1 HYPER, and then HYPER 1000 10 80 x returns $P(x)$. Pooling the results for the two days gives:

| | | Frequency | |
x	$P(x)$	O = Observed	E = Expected $= 100P(x)$
0	.4327	32	43.27
1	.3800	49	38.00
2	.1481	15	14.81
3	.0337	3	3.37
4 or more	.0055	1	.55
	1.0000	100	100.00

Grouping the last three cells to keep each $E \geq 5$ gives:

$$\chi^2 = \sum \frac{(O - E)^2}{E} = 6.12$$

Since this is greater than $\chi^2_{.05} = 5.99$ for degrees of freedom = cells $- 1 = 2$, we conclude that something is wrong with the tool. For information only, the binomial probabilities are .4475, .3616, .1443, .0379, and .0087, which lead to $\chi^2 = 8.19$, an even poorer fit.

87. Sampling Tool (continued)

348, 4. From the Problem 86 data, the estimate for the fraction yellow in the reduced population is the fraction yellow in the pooled samples, or $92/8000 = .0115$. We know $k \geq 4$, since $x = 4$ was observed in one sample. Trying $k = 4$ gives $4/N \approx .0115$, so $N \approx 348$. The expected frequencies become 34.99, 42.26, 18.82, 3.67 and .26, which lead to $\chi^2 = \Sigma(O - E)^2/E = 1.95$ with the same cell grouping as before. Because we fixed k/N to agree with the observed data, we lose another degree of freedom, so we're down to just one, for which $\chi^2_{.05} = 3.84$. No significant difference!

Increasing k and N, while keeping k/N at .0115, gives a poorer fit.

88. Marked Fish

1000. Problem 88 is Problem 87 in disguise. Both ask for the population size N based on the sample fraction occurring x/n and population occurrences k. We know:

$$\frac{x}{n} \simeq \frac{k}{N} \Rightarrow N \simeq \frac{kn}{x}$$

Here $k = 100$, $n = 50$, and $x = 5$, so $N \simeq 100(50)/5$, or 1000.

89. Sampling As You Go Along

Consider the kth unit, $k \leq n$. The ith unit will not replace it if either $iu \leq k - 1$ or $k < iu$. An equivalent condition is either $u \leq (k - 1)/i$ or $k/i < u$, and the probability of this is

$$\frac{k - 1}{i} + 1 - \frac{k}{i} = \frac{i - 1}{i}$$

The probability that the kth unit is not replaced by the ith unit as i goes from $n + 1$ through N is

$$\frac{n}{n + 1}\left(\frac{n + 1}{n + 2}\right)\left(\frac{n + 2}{n + 3}\right) \cdots \left(\frac{N - 1}{N}\right) = \frac{n}{N}$$

This is the correct, anticipated probability that a unit is in the sample.

90. Sampling As You Go Along (*continued*)

In order for the kth unit, $k > n$, to be in the sample, it must replace one of the n units in the "holding area" and not be replaced later. It gets into the holding area if $ku \leq n$ or, equivalently, $u \leq n/k$. The probability of this is n/k. The Problem 89 solution shows how to find the probability that the kth unit is not replaced by the ith unit as i goes from $n + 1$ through N. Here i goes from $k + 1$ through N, which leads to the overall probability that the kth unit makes the sample:

$$\frac{n}{k}\left(\frac{k}{k + 1}\right)\left(\frac{k + 1}{k + 2}\right) \cdots \left(\frac{N - 1}{N}\right) = \frac{n}{N}$$

91. Side-by-Side

.01. The number of combinations of n things taken r at a time is $C(n,r) = n!/[r!(n - r)!]$. Here the total number of equally likely ways to pick the two positions for the defectives is $C(200,2) = 19,900$. For the defectives to be side-by-side, the first defective can be in any one of the first 199 positions, and the second defective must immediately follow. This gives 199 arrangements, so the required probability is $199/19,900 = .01$. Since this is less than the usual .05 critical value, we conclude that the defectives do not occur at random.

92. Randomness Test for 0s and 1s

.0004. General formulas have been derived for the probability of u runs when there are m elements of one kind and n of another. See Hogg and Craig (10). If k is a positive integer like 1, 2, . . . , and $P(u)$ is the probability of u runs, we have

$$P(2k) = 2C(m - 1,k - 1)C(n - 1,k - 1)/C(m + n,m)$$
$$P(2k + 1) = [C(m - 1,k)C(n - 1,k - 1) +$$
$$C(m - 1,k - 1)C(n - 1,k)]/C(m + n,m)$$

These formulas may appear formidable, but an APL program could easily be written to do the job, since $C(n,r)$ is just $r!n$. Here $m = 10$, $n = 190$, the observed number of runs is 13, and the probability of 13 runs or less is found by summing the probabilities from $P(2)$ through $P(13)$, which gives .0004.

93. Runs Distribution

No. The maximum value of u is 21, which occurs when each of the 10 defectives is preceded and followed by a nondefective. Calculations give:

u	21	20	19	18	17	16 or less
$P(u)$	.560	.062	.282	.028	.056	.012

Whenever $m \ll n$ or vice versa, the maximum u has a high chance of happening, and the u distribution is nonnormal, like the one here. Note the second peak at $u = 19$, and $P(18) < P(17)$.

94. Randomness Test for Normal Distribution

Yes. In general, let x be a normally distributed variable ordered by time, and let n be the sample size. Then we can test the time series for randomness with the standardized mean square successive difference:

$$\frac{\sum_{i=1}^{n-1} (x_{i+1} - x_i)^2}{\sum_{i=1}^{n} (x_i - \bar{x})^2}$$

where $\bar{x} = (\Sigma x)/n$ is the sample mean. If this is within tabled limits, the data can be treated as random. Here the calculated value is 3.16, above the upper .05 probability point of 3.11 for a sample size of $n = 6$. This means that x shows too many ups and downs. A calculated value below the lower .05 probability point of .89 for $n = 6$ would mean x follows some curve. Regardless of n, the limits are equidistant from the mean of 2. See Crow, Davis, and Maxfield (2), pp. 62–64.

95. Components of Variance

67%. This is a components of variance analysis. Here we have Model C in Duncan (5), pp. 770–771. The variance of the lot means can be estimated as the variance of the sample means *minus* the portion of this variance that can be attributed just to chance sampling:

$$s_1^2 - \frac{s_2^2}{m} = 6.688 - \frac{2.933}{4} = 5.955$$

The total variance estimate is this value plus s_2^2, or $5.955 + 2.933 = 8.888$. Dividing 5.955 by 8.888 gives the 67%.

96. Components of Variance (*continued*)

5.1. If the lot means are the same, then the only variation is within-lot, and the three-sigma for x can be estimated as

$$3s_2 = 3\sqrt{2.933} = 5.1$$

A summary table is helpful:

Component	Variance	Percent	Three-sigma
Lot-to-lot	5.955	67	7.3
Within-lot	2.933	33	5.1
Total	8.888	100	8.9

This has been an example of *two-stage sampling*: lots, and then units from lots. In general, the M and N population sizes may not be indefinitely large. See Cochran (1) for finite population corrections.

97. Balls in Cells

4. This is a variation on the classical balls-in-cells problems. In effect, the inspector places $r = 32$ balls at random in $n = 20$ cells. Let x_i be 0 if the ith cell is occupied, and 1 if it is not. Then with y for the number of empty cells, and μ_y for the mean y, we have

$$y = x_1 + x_2 + \ldots + x_n$$

$$\mu_y = \mu_1 + \mu_2 + \ldots + \mu_n$$

Note the x_i are not independent, but the mean of a sum is the sum of the means, whether the variables are independent or not. The μ_i are all the same:

$$\mu_i = [0 \cdot P(x_i = 0)] + [1 \cdot P(x_i = 1)] = P(x_i = 1) = \left(\frac{n-1}{n}\right)^r$$

$$\mu_y = n\left(\frac{n-1}{n}\right)^r$$

For $r = 32$ and $n = 20$, this gives $\mu_y = 4$.

98. Poll Before Election

.28. Think of placing 100 balls into 4 cells, but we are only interested in Adams' and Carr's cells. This makes it a binomial distribution model with n equal to the 70 people who voted for Adams or Carr, x the number voting for Carr, and p equal to .5 under the hypothesis of no difference between Adams and Carr. The expected value of x is $\mu_x = np = 70(.5) = 35.0$, and the standard deviation of x is $\sigma_x = \sqrt{np(1 - p)} = \sqrt{70(.5)(.5)} = 4.1833$. The normal distribution approximation will do quite well here, since $np \gg 5$. With the continuity correction to get all of the $x = 40$ rectangle, the observed x of 40 becomes 39.5, and the corresponding z is

$$z = \frac{x - \mu_x}{\sigma_x} = \frac{39.5 - 35.0}{4.1833} = 1.08$$

The two-tail probability is .28, much greater than the usual .05 required for a significant difference.

99. Opinion Survey

Yes. Use Analysis of Means (ANOM), a modified control chart technique which takes into account the number of points being compared. See Ott (13), Chapter 5, for discussion and examples. If p is the fraction favorable for an organization of size n, then the *number* of favorable responses is pn rounded to the nearest integer, and $\bar{p} =$ favorable responses/total $= 983/1417 = .694$. Plot the data with the variable width decision limits:

$$\bar{p} \pm 2.29 \sqrt{\frac{\bar{p}(1 - \bar{p})}{n}} = .694 \pm 2.29 \sqrt{\frac{.694(.306)}{n}}$$

The 64% and 79% are outside their limits, so the differences are significant. All points must be within the limits for no significant difference.

100. Error for Survey

$\sqrt{(1/n) - (1/N)}$. See Duncan (5), pp. 572–575. If p' is the population proportion, then the required difference is $z = 1.96$ standard deviations for the hypergeometric distribution:

$$1.96 \sqrt{\frac{p'(1 - p')}{n}} \sqrt{1 - \frac{n}{N}}$$

This is a maximum when $p' = .5$, and if we replace the 1.96 with 2, we have:

$$2\sqrt{\frac{.5(.5)}{n}} \sqrt{1 - \frac{n}{N}} = \sqrt{\frac{1}{n}} \sqrt{1 - \frac{n}{N}} = \sqrt{\frac{1}{n} - \frac{1}{N}}$$

101. Sample Size for Survey

385. See Duncan (5), pp. 572–575. If p' is the population proportion, then the maximum error in the sample proportion, with 95% confidence, is $z = 1.96$ standard deviations for the hypergeometric distribution:

$$E = 1.96 \sqrt{\frac{p'(1 - p')}{n}} \sqrt{1 - \frac{n}{N}}$$

This is a maximum when $p' = .5$, and if we replace the 1.96 with 2, we have

$$E = 2\sqrt{\frac{.5(.5)}{n}} \sqrt{1 - \frac{n}{N}} = \sqrt{\frac{1}{n}} \sqrt{1 - \frac{n}{N}} = \sqrt{\frac{1}{n} - \frac{1}{N}}$$

Solving for n gives

$$n = \frac{1}{E^2 + \dfrac{1}{N}}$$

$E = .05$ and $N = 10{,}000$ gives $n = 385$. $N = \infty$ gives the easy-to-remember $n = (1/E)^2$, which is 400 for $E = .05$.

102. Nonresponse in Survey

18.0%, 98.0%. Cochran (1) deals with the problem of nonresponse in surveys. We cannot just treat the responders as a random sample from the population, because responders may tend to feel one way, and nonresponders another. Imagine the population divided into two strata: responders and nonresponders. Then assume the nonresponders would all say "no" = 0 to get the lower limit, and assume the nonresponders would all say "yes" = 1 to get the upper limit. For the proportion "yes" in the whole population, we have a lower limit of

$$\frac{(200)(.9)(1) + (1000 - 200)(0)}{1000} = .2(.9) + .8(0) = .180$$

and an upper limit of

$$\frac{(200)(.9)(1) + (1000 - 200)(1)}{1000} = .2(.9) + .8(1) = .980$$

These limits are weighted averages of the strata proportions "yes," with the weight for a stratum being the proportion of the population in the stratum.

103. Nonresponse in Survey (*continued*)

17.2%, 98.8%. First, calculate 95% confidence limits for the responders with the usual normal distribution approximation to the binomial distribution:

$$p \pm 2 \sqrt{\frac{p(1 - p)}{n}} = .900 \pm 2 \sqrt{\frac{.900(.100)}{200}}$$
$$= .900 \pm .042 = .858, .942$$

Now go back to the Problem 102 solution and use these limits in the weighted averages instead of the observed .9. The lower limit is

$$.2(.858) + .8(0) = .172$$

and the upper limit is

$$.2(.942) + .8(1) = .988$$

The limits would be even farther apart if we took into account the possible error in the .2. Problems 102 and 103 show the importance of callbacks in surveys to get as close as possible to 100% response.

104. Percent Out-of-Spec

8.1%. To estimate the percent defective, use a table of the standardized normal variable $z = (x - \mu_x)/\sigma_x$:

$$z_L = \frac{LSL - \mu_x}{\sigma_x} = \frac{150 - 161.2}{5.6} = -2.00 \Rightarrow 2.28\%$$

$$z_u = \frac{USL - \mu_x}{\sigma_x} = \frac{170 - 161.2}{5.6} = +1.57 \Rightarrow \frac{5.82\%}{8.10\%}$$

Nearly all the x values will be within three standard deviations from the mean, which is the interval $\mu_x \pm 3\sigma_x = 161.2 \pm 16.8$, or 144.4 to 178.0. The distribution is:

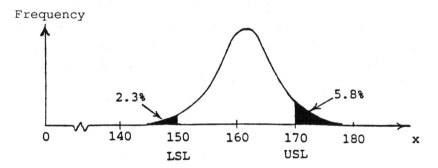

105. Moving the Mean

7.3%. The new z values with $\mu_x = 160$ are ± 1.79, which gives 3.67% in each tail. Doubling this gives the 7.3%, not much different from the 8.1%. The process is not compatible with the spec, since $6\sigma_x = 6(5.6) = 33.6$ is greater than the spec width of $170 - 150 = 20$. As long as $\sigma_x = 5.6$, the 7.3% is the best that can be done.

106. Adding and Subtracting Variables

$\mu_y = 32.38$, $\sigma_y = .50$. This is a statistical dimensioning problem that can be solved with propagation of error formulas. First of all, the mean of a sum or difference is the sum or difference of the means. Independence is not required. Further, the variance of a sum or difference is the sum of the variances, *if* the variables are independent. The variance is just the standard deviation squared. Here we have

$$\mu_y = \mu_1 + \mu_2 - \mu_3 = 39.06 + 5.64 - 12.32 = 32.38$$

$$\sigma_y = \sqrt{\sigma_1^2 + \sigma_2^2 + \sigma_3^2} = \sqrt{(.42)^2 + (.23)^2 + (.16)^2} = .50$$

Now the population percent defective for y could be found as in problems 104 and 105, if we had the specification limits.

107. Defect Density to Achieve Percent Defective

.0050. Let d be the defect density, A the area of the unit, x the number of defects on a unit, and p the fraction defective units. For random defects, x follows a Poisson distribution with mean dA, and the probability that $x = 0$ is

$$e^{-dA} = 1 - p$$

$$d = \frac{-\ln(1 - p)}{A}$$

With $p = .01$ and $A = 2$, we have $d = .0050$.

108. Percent Defective Stable?

Yes. Set up a p-chart, with p equal to the sample fraction defective. The central line is at $\bar{p} = $ defectives/units $= 45/4000 = .011$, and three-sigma control limits with sample size n are

$$\bar{p} \pm 3\sqrt{\frac{\bar{p}(1-\bar{p})}{n}} = 0.11 \pm 3\sqrt{\frac{.011(.989)}{200}}$$

This gives no lower limit and an upper limit of .033. A plot of the *percent* defective shows the points are "in control."

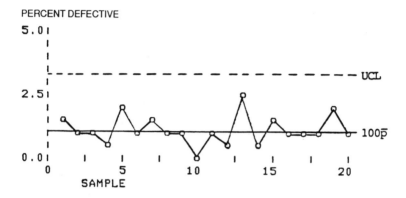

109. Defects per Unit Area Stable?

.52 ± .34. It is well known that c follows a Poisson distribution, with the mean estimated as $\bar{c}$. Since the standard deviation for a Poisson distribution is the square root of the mean, three-sigma control limits for c are

$$\bar{c} \pm 3\sqrt{\bar{c}}$$

The new variable is c/A. Dividing a variable by a constant divides the mean and standard deviation by the same constant. Hence, c/A has mean $\bar{c}/A$ and standard deviation $\sqrt{\bar{c}}/A$. The three-sigma control limits for c/A are

$$\frac{\bar{c}}{A} \pm \frac{3\sqrt{\bar{c}}}{A} = \frac{\bar{c}}{A} \pm 3\sqrt{\frac{\dfrac{\bar{c}}{A}}{A}}$$

For $A = 40$ and $\bar{c}/A = .52$, this gives .52 ± .34.

110. Standard Deviation Stable?

Yes. See Duncan (5) or Grant and Leavenworth (9). The upper
control limit (UCL) and lower control limit (LCL) for R are

$$\text{UCL} = D_4\overline{R} = 2.282(3.665) = 8.36$$

$$\text{LCL} = D_3\overline{R} = 0(3.665) = 0.00$$

These three-sigma limits for the R distribution allow for chance
sampling fluctuations only. If the R values are within the limits,
as they are here, then the interpretation is that σ_x is stable.

111. Mean Stable?

No. The three-sigma control limits for the $\overline{x}$ distribution are

$$\text{UCL} = \overline{\overline{x}} + A_2\overline{R} = 10.79 + .729(3.665) = 13.46$$

$$\text{LCL} = \overline{\overline{x}} - A_2\overline{R} = 10.79 - .729(3.665) = 8.12$$

If the $\overline{x}$ values are within these limits, then the interpretation is
that only chance sampling fluctuations are at work, and μ_x is
stable. Here we have points outside the limits, so we conclude
μ_x is not stable.

112. Process Meeting Spec?

Yes. See Duncan's acceptance control chart (5), Grant and Leavenworth's modified control limits (9), or my "Modified Control
Limits," *Quality Progress*, January 1989, pp. 44–48. We first
calculate $6\sigma_x$, the process spread, with σ_x estimated as $\overline{R}/d_2$. This
assumes the R chart shows stability. If $6\sigma_x <$ USL $-$ LSL, then
the x distribution can fit within the spec, and we can use modified
control limits on the $\overline{x}$ chart. These limits are "modified" to take
into account the spec. They allow μ_x to move around, as long as
$\mu_x \pm 3\sigma_x$ stays within LSL and USL. Hence, the upper modified
control limit (UMCL) is the classical UCL for a process mean at
USL $- 3\sigma_x$:

$$\text{UMCL} = (\text{USL} - 3\sigma_x) + A_2\overline{R}$$

Since $A_2\overline{R}$ is a disguised way of writing $3\sigma_{\overline{x}}$ or $3\sigma_x/\sqrt{n}$ this substitution and factoring gives

$$\text{UMCL} = \text{USL} - \left(3 - \frac{3}{\sqrt{n}}\right)\sigma_x$$

Similar reasoning gives the lower modified control limit (LMCL):

$$\text{LMCL} = \text{LSL} + \left(3 - \frac{3}{\sqrt{n}}\right)\sigma_x$$

Here the R-chart shows stability, and we have

$$\sigma_x = \frac{\overline{R}}{d_2} = \frac{3.665}{2.059} = 1.78$$

Since $6\sigma_x = 6(1.78) = 10.68$ is less than the spec width of USL $-$ LSL $= 20 - 2 = 18$, we can use modified control limits. With USL $= 20$, LSL $= 2$, $n = 4$, and $\sigma_x = 1.78$, we find UMCL $= 17.33$ and LMCL $= 4.67$. All the $\overline{x}$ values meet these limits, so we conclude that the process is meeting the spec.

113. *p*-Chart ARL Equals Cusum ARL

$c = [k] = [h + k]$. Suppose we are starting the Cusum plan. Then we accept and start a new sum if

$$x - k \leq 0 \Rightarrow x \leq k \Rightarrow x \leq [k],$$

where $[k]$ is the greatest integer in k. Hence, if $c = [k]$, the condition for acceptance and forgetting prior history is the same for p-Chart and Cusum. We reject on the first cusum if

$$x - k > h \Rightarrow x > h + k \Rightarrow x \geq [h + k] + 1$$

If $c = [h + k]$, then the condition for rejection is the same for p-Chart and Cusum. Putting it all together, we see that p-Chart and Cusum operate exactly the same way, with no use of prior history, if

$$c = [k] = [h + k]$$

This means that given a p-Chart plan, we can always find an equivalent Cusum plan by letting $k = c$ and $h = .05$. This is one of the many (h, k) combinations that will do the job. On the other hand, given a Cusum plan, we can find an equivalent p-Chart plan *only* if $[k] = [h + k]$, and that common value is the required c.

114. p-Chart ARL Greater than Cusum ARL

$[k] \neq [h + k] \leq c$. For any sample, the Cusum acceptance number (maximum number of defectives allowed) depends on the prior cusum. The loosest (highest) possible acceptance number happens when the prior cusum is zero, because then we are farthest from the h limit. We accept if

$$x - k \leq h \Rightarrow x \leq h + k \Rightarrow x \leq [h + k]$$

Suppose $[k] \neq [h + k]$, so there are some cusums greater than zero. Then the Cusum acceptance number won't be $[h + k]$ all the time. Hence, if $[k] \neq [h + k] \leq c$, the p-Chart must give fewer rejections than Cusum, and the p-Chart curve lies above the Cusum curve.

115. p-Chart ARL Less than Cusum ARL

$c \leq [k] \neq [h + k]$. The tightest (smallest) Cusum acceptance number happens when the prior cusum is just under h. In order to have any cusum other than zero, we must have $[k] \neq [h + k]$. If the prior cusum is just under h, then in order to *stay* under, we must have

$$x - k \leq 0 \Rightarrow x \leq k \Rightarrow x \leq [k]$$

The Cusum acceptance number won't be $[k]$ all the time. Hence, if $c \leq [k] \neq [h + k]$, the p-Chart must give more rejections than Cusum, and the p-Chart curve lies below the Cusum curve.

116. ARL for Control Limits

370. Consider an indefinitely long string of $\bar{x}$ points from the process. Let p be the probability that a single randomly selected

point is outside the three-sigma control limits, and let r_i be the *i*th run length. Then we have

$$\text{ARL} = \lim_{n \to \infty} \frac{r_1 + r_2 + \ldots + r_n}{n} = \frac{\text{Total points}}{\text{Signals}} = \frac{1}{p}$$

Duncan's normal distribution table gives $p = 2(1 - .9987) = .0026$, and ARL $= 1/.0026$, or 385, but a more accurate value can be found with APL. Execute)LOAD 42 STAT2, and then NDTR Z returns the cumulative probability for the standardized normal variable Z. Here we find p with $2 \times 1 - \text{NDTR } 3$, which is .002699 . . . The ARL is $\div 2 \times 1 - \text{NDTR } 3$, which is 370.379. . . .

117. ARL for Rule of Seven

127. As in the Problem 116 solution, consider an indefinitely long string of $\bar{x}$ points from the process. If we let p be the probability that a single randomly selected point completes a run of seven on one side of the target, then the required ARL is $1/p$. Consider a high run. Let H stand for a high point, L for a low point, and X for anything except a nonsignal H. This means X can be a high-signal point or a low point (signal or not). Then in order to be a signal point for a high run, a point must be the last point in this sequence of X and seven Hs: $XHH \ldots H$. Let p_H be the probability of this, so p_H is the product of the probabilities for the eight independent events. The probability of H is $P(H) = .5 = P(L)$. The probability of X is the sum of the probabilities for the two mutually exclusive events, high-signal or low: $P(X) = p_H + .5$. This leads to

$$p_H = (p_H + .5)(.5)^7 \Rightarrow p_H = \frac{(.5)^8}{1 - (.5)^7}$$

The p_L probability for a low run must be the same, and the low and high runs are mutually exclusive events, so we have: $p = p_L + p_H = 2p_H$. ARL $= 1/p$ reduces to $2^7 - 1$. This agrees with a formula in Feller (6) for the ARL for a success run of length r or a failure run of length ρ, with p = P(success) and q = P(failure) = $1 - p$:

$$\text{ARL} = \frac{(1 - p^r)(1 - q^\rho)}{qp^r + pq^\rho - p^rq^\rho}$$

When $r = \rho$, and $p = q = .5$, this reduces to $2^r - 1$.

118. ARL for Control Limits and Rule of Seven

96. Modify the solution to Problem 117 as follows. If a point completes a run of seven *and* is outside the control limits, then classify it as a signal for outside the control limits. This makes the ways to get a signal mutually exclusive, so $p = p_L + p_H + .002699$ . . . For the high run, let H stand for a point between the target and the upper control limit, L for a point between the target and the lower control limit, and X for anything except a nonsignal H. This means X can be a high-signal point for a run, or a low point between the target and lower control limit (signal or not), or a point outside the control limits. From APL, $P(H) = P(L) = .498650$ . . . , and $P(X) = p_H + .498650$ . . . $+ .002699$ . . . We have, then,

$$p_H = (p_H + .498650 . . .$$

$$+ .002699 . . .)(.498650 . . .)^7 \Rightarrow p_H = .003873 . . .$$

The p_L probability for a low run must be the same, so $p = 2(.003873 . . .) + .002699 . . . = .010446$ . . . , and ARL = $1/.010446$ . . . $= 95.729$. . . . For a Markov chain approach to this problem, see C. W. Champ and W. H. Woodall's "Exact Results for Shewhart Control Charts With Supplementary Runs Rules," *Technometrics*, November 1987, pp. 393–399. For a point outside the control limits or a run of *eight* points on one side of the target, Champ and Woodall give ARL = 152.73, which can be reproduced with the method given here. They do not consider the Rule of Seven.

119. Measurements on a Standard

It can't be done with these readings! Plot the $n = 8$ readings versus time, and you will see that the readings trend up, and then down, so the tester needs some maintenance work. In general, to test a time series for random variation from a normal distribution, calculate the standardized mean square successive difference for the variable x:

$$\frac{\sum_{i=1}^{n-1} (x_{i+1} - x_i)^2}{\sum_{i=1}^{n} (x_i - \bar{x})^2},$$

where $\bar{x} = (\Sigma x)/n$ is the sample mean. If this is within tabled limits, the data can be treated as a random sample. Here the calculated value is .88, below the lower .05 probability point of .98 for a sample size of 8. See Crow, Davis, and Maxfield (2), pp. 62–64.

120. Standard Value and Error

$26.03 \pm .11$. The population here is an indefinitely large number of readings on the standard, and the true value is the population mean μ_x. For an estimate, use the sample mean $\bar{x}$, with population standard deviation $\sigma_{\bar{x}} = \sigma_x/\sqrt{n} \simeq s_x/\sqrt{n}$, where n is the sample size, and s_x is the sample standard deviation. We have then

$$\bar{x} \pm \frac{3s_x}{\sqrt{n}} = 26.03 \pm \frac{.311}{\sqrt{8}} = 26.03 \pm .11$$

121. Tester Bias

$.15 \pm .34$. For sample size n and population standard deviation σ_x, the line tester's long-run mean is

$$\bar{x} \pm \frac{3\sigma_x}{\sqrt{n}} \simeq \bar{x} \pm \frac{3s_x}{\sqrt{n}}$$

Here this turns out to be $26.18 \pm .32$. A well-known propagation of error formula gives the error in the sum or difference of two independent variables as the square root of the sum of the squares of the individual errors. Hence, the bias estimate with its three-sigma error is

$$(26.18 - 26.03) \pm \sqrt{(.32)^2 + (.11)^2} = .15 \pm .34$$

Since zero is in the interval, the observed difference in means is not significant, and no tester adjustment is necessary. Note the maximum error in a single reading due to repeatibility is $\pm 3\sigma_x \simeq \pm 3s_x$, or ± 1.43.

122. Missing Defects

5. This is a binomial distribution model with $n = 25$ defects, $p = .10$ for the long-run fraction missed, and $x =$ number of misses. We want to find c such that the probability of c or fewer misses is at least .95. The probabilities can be found with tables, a computer program, or the easy-to-use monograph in H. R. Larson's "A Nomograph of the Cumulative Binomial Distribution," *Industrial Quality Control*, December 1966, pp. 270–278. We find $P(x \leq 4) = .90$ and $P(x \leq 5) = .97$, so $c = 5$ is the required answer. Note the expected number of misses is $np = 2.5$, and we allow 5 for chance variation above the 2.5.

123. Measurement of Variable on Same Units

$0 \pm .385$. A well-known propagation of error formula gives the error in the sum or difference of two independent variables as the square root of the sum of the squares of the individual errors, so the required limits are

$$0 \pm \sqrt{(.372)^2 + (.100)^2} = 0 \pm .385$$

124. Measurement of Variable on Independent Sample of Units

.14. For this problem, most statistics books offer only a *t*-test that requires that the population standard deviations be equal. This seems unreasonable here, since the *F*-test on the sample standard deviations shows a significant difference at the .05 level of significance:

$$F_{obs}(4,4) = (.79/.15)^2 = 27.7 > F_{.025}(4,4) = 9.60$$

A modified *t*-test that does *not* require equal population standard deviations is the Aspin-Welch test. See Duncan (5), pp. 616–618. We calculate

$$t = \frac{\bar{x}_1 - \bar{x}_2}{\sqrt{\dfrac{s_1^2}{n_1} + \dfrac{s_2^2}{n_2}}}$$

The degrees of freedom v is not a set $n_1 + n_2 - 2$ as it is for the equal-standard-deviations t, but instead:

$$v = \frac{1}{\dfrac{c^2}{n_1 - 1} + \dfrac{(1 - c)^2}{n_2 - 1}}, \text{ with } c = \frac{\dfrac{s_1^2}{n_1}}{\dfrac{s_1^2}{n_1} + \dfrac{s_2^2}{n_2}}$$

It will be left as an exercise for the reader to show that v is always between a maximum of $n_1 + n_2 - 2$ and a minimum equal to the smaller of $n_1 - 1$ and $n_2 - 1$. Here $t = 1.891$, $c = .034797$, and $v = 4$ rounded to the nearest integer. Interpolation in a *t*-table gives the two-tail area of .14, which is the required probability.

125. Percent Defective for Same Units

Goods		Bads	
QC = True	Maximum allowable manufacturing misses	QC = True	Maximum allowable manufacturing misses
50	2	0	—
49	2	1	1
48	2	2	1
47	2	3	1
46	2	4	2
45	2	5	2

Use the binomial distribution with n = trials equal to QC goods or QC bads, p = probability of occurrence on a trial equal to α

or β, and x = occurrences equal to misses. On APL, execute)COPY 42 STAT6 BINOCDF, and then .01 BINOCDF 50 2 returns the probability of 2 or fewer occurrences, for example. Systematic trial and error gives the table.

126. Percent Defective for Independent Sample of Units

5. For this type of problem, you could use Fisher's Exact Test for two proportions (Problem 174), but the calculations can be lengthy. A simple alternative with intuitive appeal is the Balls/ Cells approximation, which gives nearly the same results if the fraction defective for the pooled samples is less than .20. Think of the defectives as balls being placed at random into two cells, Manufacturing and Quality Control. This is a binomial distribution model with $n = x_1 + x_2$ trials and $p = .5$ probability of occurrence, with an occurrence being a ball going to the Manufacturing cell. The probability of such an unequal split as $x_2 > x_1$ or worse in that direction is given by the binomial distribution as

$$P(x_1 \text{ or less}) = \sum_{x=0}^{x_1} C(n,x)p^x q^{n-x}$$

where $q = 1 - p$, and $C(n,x) = n!/[x!(n - x)!]$ is the number of combinations of n things taken x at a time. With $x_1 = 0$, the required x_2 is the lowest x_2 such that $(.5)^{x_2} \le .05$. Taking the ln of each side gives $x_2 \ge 4.3$, or $x_2 = 5$. For $x_1 > 0$, trial and error with the binomial distribution gives the lowest x_2 such that $P(x_1 \text{ or less}) \le .05$. We end up then with minimum contrasts in x_1 and x_2 required for statistical significance when $n_1 = n_2$: (0,5), (1,7), (2,9), etc.

127. Sampling After Imperfect 100% Inspection

.22. We could use the hypergeometric distribution here, but it's more fun to use basic probability. For the first unit drawn, the probability of a nondefective is nondefectives/total, or 8/10. Then for the second, it's 7/9, etc. The overall probability of all non-defectives is the product of these values:

$$\left(\frac{8}{10}\right)\left(\frac{7}{9}\right)\left(\frac{6}{8}\right)\left(\frac{5}{7}\right)\left(\frac{4}{6}\right) = \frac{2}{9}, \text{ or } .22$$

128. Acceptable Quality Level

.85%. Let $100p$ be the required percent defective, so p is the fraction defective. In general, for population size N and $n/N <$ .10, the probability of acceptance P_a is given by the binomial distribution:

$$P_a = P(c \text{ or less}) = \sum_{x=0}^{c} C(n,x)p^x q^{n-x}$$

where $q = 1 - p$, and $C(n,x) = n!/[x!(n - x)!]$ is the number of combinations of n things taken x at a time. This is difficult to solve for p unless $c = 0$, in which case P_a reduces to $(1 - p)^n$. Setting this equal to .95 for our problem gives

$$(1 - p)^6 = .95 \Rightarrow p = 1 - .95^{\frac{1}{6}} = .0085, \text{ or } .85\%$$

129. Acceptable Quality Level vs. Lower Limit

.85%. The required limit is the percent defective so low that $P(x \geq 1) = .05$. But $P(x \geq 1) = .05$ means $P(x = 0) = .95$, and we know the percent defective for that is .85% from Problem 128. Since the probability for a point like the lower limit is zero, we conclude the week is *worse* than the AQL. In general, when $x = c + 1$ defectives are observed, the 95% confidence one-sided lower limit is the AQL, because $P(x \geq c + 1) = .05$ is exactly the same condition as $P(x \leq c) = .95$. This suggests a way to find the AQL when n and $c > 0$ are given. Use a canned program to find the 95% confidence one-sided lower limit for $x = c + 1$ defectives in n. On APL, execute)LOAD 42 STAT2, and then .90 BICI $c + 1$ n returns 90% confidence two-sided limits. Just the lower limit is the 95% confidence one-sided lower limit we want. For example, if $n = 6$ and $c = 1$, then .90 BICI 2 6 returns .0628 . . . , or 6.3% for the AQL.

130. Acceptable Quality Level vs. Significance Test

$x = c + 1 = 1$. At the AQL, $P(x \leq c) = .95$ means $P(x \geq c + 1) = .05 = \alpha$, so $x = c + 1$ is the point at which the observed percent defective becomes significantly worse than the AQL.

Problems 129 and 130 illustrate a general truth. For significance level α, a necessary and sufficient condition for "no significant difference" between a population parameter and its sample estimate is that the population parameter be within the $100(1 - \alpha)\%$ confidence interval.

131. Good and Bad Quality

$n = 98$, $c = 4$. This is the example on the binomial distribution nomograph in H. R. Larson's "A Nomograph of the Cumulative Binomial Distribution," *Industrial Quality Control*, December 1966, pp. 270–278. In general, let α be the producer's risk of rejection at p_1 fraction defective, and β be the consumer's risk of acceptance at p_2 fraction defective. Here $\alpha = .05$, $p_1 = .02$, $\beta = .10$, and $p_2 = .08$. In practice, $100p_1$ is called the acceptable quality level (AQL), and $100p_2$ the lot tolerance percent defective (LTPD). Approaches to this type of problem are:

a. Binomial distribution nomograph.
b. F. E. Grubbs' table in Duncan (5), p. 172. This uses the Poisson distribution and is a good approximation if $p_2 < .10$.
c. The APL program SAMPLINGPLAN, which is equivalent to Grubbs' table. First, execute)COPY 42 STAT6 QC. Then, for example, .02 .08 SAMPLINGPLAN .05 .10 returns $n = 116$ and $c = 5$. This differs from the nomograph solution because the program uses the Poisson distribution and keeps the α and β errors just *less* than or equal to the specified values, as opposed to the *nearest* solution.
d. Normal distribution, as described in Duncan (5), Chapter 25. This is a good approximation if $np_1 \geq 5$.

The goodness of any solution can be checked with the exact binomial distribution, after)COPY 42 STAT6 BINOCDF. For example, .02 BINOCDF 98 4 returns .953.

132. Widgets in Tubes

42%. This is a hypergeometric distribution model, with $N = 561$ total tubes, $n = 8$ tubes in the sample, and $p = 58/561 = .103$

fraction bad tubes. Since $n/N = 8/561 = .01 < .10$, the more familiar binomial distribution with $N = \infty$ is a good approximation here. The probability of acceptance is the probability of zero bad tubes, and the binomial distribution gives this as $(1 - p)^n = (1 - .103)^8 = .419$, or 42%. The more accurate hypergeometric probability is .415. By contrast, if the sample were a random 400 widgets spread in a haphazard way throughout the lot, then the binomial distribution with $n = 400$ and $p = .103$ fraction defective widgets gives an acceptance probability of virtually zero.

133. Zero Defects in Sample

4.5%. In general, let n be the sample size, N the lot size, and p the unknown lot fraction defective. The upper limit with 90% confidence is that value of p for which the probability of zero defectives is only $1 - .90$, or .10. Since $n/N < .10$ here, the binomial distribution will be a good approximation to the more accurate hypergeometric distribution. Use the nomograph in H. R. Larson's "A Nomograph of the Cumulative Binomial Distribution" (*Industrial Quality Control*, December 1966, pp. 270–278) to get $p = .045$, or 4.5%. Alternatively, set $(1 - p)^{50} = .10$, and solve for p by taking logs of both sides of the equation. Note that zero defects in the sample doesn't guarantee zero defects in the lot.

134. LTPD and Upper Limit

6.5%. An accepted lot has at most 2 defectives in the sample. The required upper limit is that value for which the probability of 2 or less is 100% minus the 90% confidence, or 10%. This is the LTPD definition!

135. Average Outgoing Quality for Sampling Plan

.16%. For k lots, with k indefinitely large, we have

$$\text{AOQ} = 100 \, \frac{\text{Defectives shipped}}{\text{Total units shipped}}$$

The denominator is easy; it's just kN. The numerator is a little harder. The only defectives shipped are the ones in the uninspected portions of the accepted lots. In general, let p be the process fraction defective and P_a be the probability of acceptance for a lot. Then $P_a k$ lots are accepted, and each of these has $p(N - n)$ defectives, on the average. Hence, we have

$$AOQ = 100 \frac{[P_a k][p(N - n)]}{kN} = P_a(100p)\left(1 - \frac{n}{N}\right)$$

P_a is given by the binomial distribution:

$$P_a = P(c \text{ or less}) = \sum_{x=0}^{c} C(n,x) \, p^x q^{n-x}$$

where $q = 1 - p$, and $C(n,x) = n!/[x!(n - x)!]$ is the number of combinations of n things taken x at a time. Here $N = 400$, $n = 50$, $c = 0$, and $100p = .20\%$, which gives $P_a = .9047$ and $AOQ = .16\%$.

136. Average Outgoing Quality Limit for Sampling Plan

.64%. When $c = 0$, P_a reduces to $(1 - p)^n$, so we have

$$AOQ = (1 - p)^n (100p)\left(1 - \frac{n}{N}\right)$$

Setting the first derivative of AOQ with respect to p equal to zero and solving for p gives the p at which the AOQ is a maximum:

$$100\left(1 - \frac{n}{N}\right)[(1 - p)^n + pn(1 - p)^{n-1}(-1)] = 0$$

$$(1 - p) - np = 0$$

$$p = \frac{1}{n + 1}$$

Substituting this value of p in the AOQ equation gives the AOQL:

$$AOQL = 100\left(\frac{n}{n + 1}\right)^n \left(\frac{1}{n + 1}\right)\left(1 - \frac{n}{N}\right)$$

$N = 400$ and $n = 50$ gives AOQL $= .64\%$. Dodge and Romig's formula gives the same thing.

137. Average Outgoing Quality for Imperfect 100% Inspection

.5%. This is an example from my "Sampling Plan Adjustment for Inspection Error and Skip-Lot Plan" (*Journal of Quality Technology*, July 1982, pp. 105–116). For N inspected units, let p' be the true initial fraction defective, and p be the expected value of the observed fraction defective. Observed defectives are either true bads correctly called bad, or true goods incorrectly called bad, so the relation between p and p' is

$$p = \frac{p'(1 - \beta)N + (1 - p')\alpha N}{N} = \alpha + (1 - \alpha - \beta)p'$$

$$p' = \frac{p - \alpha}{1 - \alpha - \beta}$$

For $p = .05$, $\alpha = .01$, and $\beta = .10$, we have $p' = .045$. After the screening, defectives shipped are units truly defective *and* missed, or $p'\beta N$. The N is reduced by pN, so we have:

$$\text{AOQ} = 100\,\frac{p'\beta N}{N - pN} = \frac{100p'\beta}{1 - p} = .5\%.$$

138. AQL and AOQL Requirements

$n = 80, c = 2$. For $c = 0, 1, 2, \ldots$, use the binomial distribution to find n for a 95% chance of acceptance at the AQL of 1%. This could be done with a computer program or H. R. Larson's nomograph reproduced in my "Sampling Plan Adjustment for Inspection Error and Skip-Lot Plan" (*Journal of Quality Technology*, July 1982, pp. 105–116). For an approximation, use F. E. Grubbs' Poisson distribution table in Duncan (5), p. 172. Next, find the AOQL for each plan from

$$\text{AOQL} = \frac{100y}{n}\left(1 - \frac{n}{N}\right)$$

where y is a factor from Duncan (5), p.376. Here the nomograph and AOQL formula give

c	n	AOQL(%)
0	4	9.2
1	35	2.4
2	80	1.6
3	135	1.3

139. Variables Plan

$n = 32, \bar{x} \leq 50.7$. See Duncan (5), Chapter 11. Let z_1 and z_2 be the standardized normal variables for right-hand tail areas of p_1 and p_2. Here $z_1 = 3.090$ and $z_2 = 2.576$. For the $\alpha = .05$ and $\beta = .10$ sampling risks, we have $z_\alpha = 1.645$ and $z_\beta = 1.282$. The sample size, then, is

$$n = \left(\frac{z_\alpha + z_\beta}{z_1 - z_2} \right)^2$$

The general procedure when there is an upper spec limit U is to accept if $\bar{x} \leq U - k\sigma_x$. Duncan gives two k formulas: one for fixing p_1 and the other for fixing p_2. Substituting the n expression in these formulas shows that they both reduce to

$$k = \frac{z_\alpha z_2 + z_\beta z_1}{z_\alpha + z_\beta}$$

This formula is given in Duncan (5), Chapter 12. Calculations here give $n = 32$ and $k = 2.801$, so we test 32 units and accept if $\bar{x} \leq 65 - 2.801(5.1)$, or $\bar{x} \leq 50.7$. Of course, the s_x chart must be continued, as proof that we "know" σ_x.

140. Sequential Plan

562. See Duncan (5) and Crow, Davis, and Maxfield (2). The inspection results can be plotted on a graph with the horizontal axis n = number of units inspected, and the vertical axis d = number of defectives found. The parallel decision lines are

$$d = -h_1 + sn, \text{ for acceptance}$$

$$d = h_2 + sn, \text{ for rejection}$$

Both references give the equations for h_1, h_2, and s. As we inspect, we take a walk over the (n,d) plane. After each unit is inspected, we do one of the following:

a. Accept if we are at or below the accept line.
b. Reject if we are at or above the reject line.
c. Continue sampling if we are between the lines.

Crow et al. (2) give a truncation value for n, at which point we accept if we are closer to the accept line, or reject if we are closer to the reject line. Here we have $h_1 = 1.3953$, $h_2 = 1.7914$, and $s = .002487$. The minimum inspection occurs when we have 0 defectives and pass into the accept region:

$$0 \le -1.3953 + .002487n$$

This gives $n \ge 561.04$, so n must be 562.

141. Average Sample Size

831. Duncan (5) gives formulas for the average sample size as a function of the incoming quality. For reasons of propriety, this average sample size is called the average sample *number*, or ASN. At $100p_1$, with the chance of acceptance equal to $1 - \alpha$, we have

$$\text{ASN} = \frac{(1 - \alpha)h_1 - \alpha h_2}{s - p_1}$$

For the present problem, with $\alpha = .05$, this gives ASN = 831.

142. Sample vs. Screen

.19. This is a finite population problem with population size $N = 300$, population defectives $k = 5 + 7 = 12$, Manufacturing sample size $n = 220$, and Manufacturing sample defectives $x = 7$. According to the hypergeometric distribution, the probability of exactly x defectives is

$$P(x) = \frac{[C(k,x)][C(N - k, n - x)]}{C(N,n)}$$

where $C(n,r) = n!/[r!(n - r)!]$ is the number of combinations of n things taken r at a time. For an APL program, execute)COPY

42 STAT1 HYPER, and then HYPER 300 12 220 7 returns .19 for the probability of 7 or less in the 220. The program sums $P(x)$ from $x = 0$ through $x = 7$. Since the probability here is greater than the usual .05 critical value, we would say the difference in the two inspections is not statistically significant—that is, the difference can be reasonably attributed just to chance.

143. Average Run Length

365. In general, suppose we have repeated trials with a probability p for an occurrence on each trial. The mean number of trials to an occurrence is called the ARL = average run length. If r_i is the ith run length, then we have

$$\text{ARL} = \lim_{n\to\infty} \frac{r_1 + r_2 + \ldots + r_n}{n} = \frac{\text{Trials}}{\text{occurrences}} = \frac{1}{p}$$

For an infinite series proof, see Duncan (5), pp. 433–434. Here $p = 1/365$, so ARL = 365. As another example, if we flip a coin until a head appears, then the median waiting time is 1, and the mean waiting time is 2. It can be shown that the median is always less then the mean, regardless of the p value.

144. Average Run Length and Upper Limit

1.7, 3. In Problem 143, we saw that the ARL for an event is the reciprocal of the probability of occurrence, so here we have ARL = $1/(1 - .41) = 1.7$. Let x be the variable run length. Then a 90% or more chance that $x \le L$ means the chance that we accept the first L lots must be 10% or less. It follows that L is the lowest integer x such that $P_a^x \le .10$, or $x \ge (\ln.10)/\ln P_a$. Here $(\ln.10)/(\ln.41) = 2.6$, so $L = 3$. Hence, on the average we reject in $1.7 \cong 2$ lots, and 90% of the time in 3 or less.

145. Percent Defective for Accepted Lots

.35%. This one is easy if and only if you read the *Journal of Quality Technology*; see G. J. Hahn's "Estimating the Percent Nonconforming in the Accepted Product After Zero Defect Sampling," July 1986, pp.182–188. For the required estimate $\hat{p}$, Hahn proposes the following procedure:

$$A = \frac{\text{Number of lots with } x = 1}{n}$$

$$B = \text{Number of lots with } x = 0$$

$$\hat{p} = 100\left(\frac{A}{A + B}\right)$$

Here $A = 7/200$, $B = 10$, so $\hat{p} = .35\%$. The x values were generated at random on APL, with 25 of the lots at .35% defective, and the other 5 at 3.0% defective. All the 3.0% lots were rejected, so the true quality of the accepted lots *is* .35%! See Hahn's paper for the rationale behind this amazing "jackknife" procedure and a discussion of other estimates.

146. Breakeven Fraction Defective

.010. Let N be the lot size. At p_b, the screening loss is the inspection cost NI, plus the scrap cost $(p_bN)C$. The acceptance loss is $(p_bN)A$. Setting these losses equal to each other gives

$$NI + p_bNC = p_bNA$$

$$p_b = \frac{1}{A - C}$$

For the cost figures here, $p_b = .010$.

147. Sampling Plan for Breakeven Fraction Defective

2 or 3, take your pick. At $p_b = .010$, we don't care whether we accept or reject, so the probability of acceptance P_a should be .50. Then lots with $p < p_b$ have $P_a > .50$, and lots with $p > p_b$ have $P_a < .50$. Hence, we have a better than 50% chance of making the right economic decision. To find the required c, see the operating characteristic curves and tabulated values in the back of Mil-Std-105D (17). For Sample Size Code Letter M and

$P_a = .50$, $c = 2$ gives $p = .0085$, and $c = 3$ gives $p = .0117$. The $p_b = .010$ is just about midway between the two p values, so $c = 2$ or $c = 3$ could be used. Problems 146 and 147 are based on Dr. John W. Enell's "Which Sampling Plan Should I Choose?", first published in 1954 in *Industrial Quality Control*, and reprinted in the *Journal of Quality Technology*, July 1984, pp. 168–171.

148. Fine-Tuning the Sampling Plan for Breakeven Fraction Defective

267. There are many approaches to this problem, but perhaps the simplest one is to use the Poisson distribution table given in J. M. Cameron's "Tables for Constructing and for Computing the Operating Characteristics of Single-Sampling Plans," *Industrial Quality Control*, July 1952, pp. 37–39. For $c = 2$ and $P_a = .50$, Cameron's Table 1 gives $np = 2.674$. Our np is $.010n$, so setting $.010n = 2.674$ and solving for n gives $n = 267$.

149. Standard Deviation

$n = 10$, $s_a = 34.3$. See Duncan (5), pp. 360–362. Let $\chi^2_{P,n-1}$ be the chi-square value for right-hand probability P and $n - 1$ degrees of freedom. Then it can be shown that for risk of rejection α at σ_1, and risk of acceptance β at σ_2, we must have

$$\frac{(n-1)s_a^2}{\sigma_1^2} = \chi^2_{\alpha,n-1} \quad \text{and} \quad \frac{(n-1)s_a^2}{\sigma_2^2} = \chi^2_{1-\beta,n-1}$$

Dividing these two equations gives

$$\left(\frac{\sigma_2}{\sigma_1}\right)^2 = \frac{\chi^2_{\alpha,n-1}}{\chi^2_{1-\beta,n-1}}$$

Here $\alpha = .05$, $\beta = .10$, and the required ratio of chi-square values is $(50/25)^2$, or 4. Trial and error in Duncan's Table C (5) shows that the nearest solution is $16.9/4.17 = 4.05$ for $n - 1 = 9$, or $n = 10$. Solving the σ_1 equation for s_a gives $s_a = 34.3$.

150. Mean When Standard Deviation Known

$n = 12$, $\bar{x}_a = .1322$. See Duncan (5), pp. 342–343. Let z_P be the z value for right-hand probability P. Then it can be shown that for risk of rejection α at μ_1, and risk of acceptance β at μ_2, we must have

$$\bar{x}_a = \mu_1 - \frac{z_\alpha \sigma_x}{\sqrt{n}} = \mu_2 + \frac{z_\beta \sigma_x}{\sqrt{n}}$$

If we were testing for too *high* a mean, rather than too low, we would have

$$\bar{x}_a = \mu_1 + \frac{z_\alpha \sigma_x}{\sqrt{n}} = \mu_2 - \frac{z_\beta \sigma_x}{\sqrt{n}}$$

In either case, solving for n gives

$$n = \left(\frac{z_\alpha + z_\beta}{|\mu_2 - \mu_1|/\sigma_x} \right)^2$$

Here $\alpha = .05$, $\beta = .10$, $z_\alpha = 1.645$, $z_\beta = 1.282$, $\mu_1 = .135$, $\mu_2 = .130$, and $\sigma_x = .006$, so $n = 12$. Using this in the $\bar{x}_a$ equation for the α requirement gives $\bar{x}_a = .1322$.

151. Mean When Standard Deviation Unknown

$n = 14$, $\bar{x}_a = .135 - 1.771 s_x/\sqrt{14}$. The exact solution for n requires the use of a double integral. See the reference under Duncan's Figure 15.9, (5), p. 356. However, a simple approximation can be derived as follows. Since we don't know σ_x, we use s_x and Student's t, with $t_{P,n-1}$ equal to the t value for right-hand probability P and $n - 1$ degrees of freedom. We want

$$\bar{x}_a = \mu_1 - \frac{t_{\alpha,n-1} s_x}{\sqrt{n}} = \mu_2 + \frac{t_{\beta,n-1} s_x}{\sqrt{n}}$$

If we were testing for too *high* a mean, rather than too low, we would have

$$\bar{x}_a = \mu_1 + \frac{t_{\alpha,n-1} s_x}{\sqrt{n}} = \mu_2 - \frac{t_{\beta,n-1} s_x}{\sqrt{n}}$$

In either case, we have

$$\frac{|\mu_2 - \mu_1|}{s_x} = \frac{t_{\alpha,n-1} + t_{\beta,n-1}}{\sqrt{n}}$$

We don't know what s_x is going to be, but we can estimate it with the historical average for s_x. Here $s_x \simeq .006$, so $|\mu_2 - \mu_1|/\sigma_x \simeq$ .005/.006, or .83. Trial and error in Duncan's Table B (5) shows that $n = 14$ gives $(t_{.05,13} + t_{.10,13})/\sqrt{14} = .834$, which is the nearest solution. The double integral approach gives .826 for $n = 14$. Finally, the α requirement gives $\bar{x}_a$.

152. Double Sampling Plan

.77. A proper double sampling plan should have a rejection number after the first sample, and the decision after the second sample should be based on all units inspected, not just the second sample. Nonetheless, the given plan *is* a decision rule, and we can find the probability of acceptance with the binomial distribution, which gives the probability of x defectives in a sample of size n units from a p fraction defective process as

$$P(x) = C(n,x)p^x q^{n-x}$$

where $q = 1 - p$, and $C(n,x) = n!/[x!(n - x)!]$ is the number of combinations of n things taken x at a time. Here the probability of acceptance P_a is

$$P_a = P(0 \text{ in } 4) + [P(1 \text{ or more in } 4)] [P(0 \text{ in } 4)]$$

Since $P(1 \text{ or more in } 4) = 1 - P(0 \text{ in } 4)$, we have

$$P_a = (1 - p)^4 + [1 - (1 - p)^4][1 - p]^4$$

Substituting $p = .15$ gives $P_a = .77$.

153. Double Sampling Plan (*continued*)

AQL $= 6\%$, LTPD $= 52\%$. The P_a can be written as

$$P_a = 2(1 - p)^4 - [(1 - p)^4]^2$$

The AQL is found by letting $x = (1 - p)^4$ and setting $P_a = .95$:

$$x^2 - 2x + .95 = 0$$

The quadratic formula then gives

$$x = \frac{2 \pm \sqrt{4 - 4(.95)}}{2} = 1 \pm .2236$$

The value we want is $x = 1 - .2236 = .7764$. Setting $x = (1 - p)^4 = .7764$ and solving for p gives $p = 1 - (.7764)^{1/4} = .0613$, or 6%. For the LTPD, proceed as above, except $P_a = .10$. This leads to $p = .5241$, or 52%.

154. Conventional Double Sampling Plan

AQL = 5%, LTPD = 47%. The only way we can get to a second sample is to have one defective in the first sample, so the probability of acceptance becomes

$$P_a = (1 - p)^4 + [4p(1 - p)^3][1 - p]^4$$
$$= (1 - p)^4 + 4p(1 - p)^7$$

The AQL is found by setting $P_a = .95$:

$$4p(1 - p)^7 + (1 - p)^4 - .95 = 0$$

There is apparently no easy way to solve this equation, other than systematic trial and error. On APL, which operates from right to left and within parentheses first, the P_a for $p = .06$ can be found by executing

$$((1 - .06)*4) + 4 \times .06 \times (1 - .06)*7$$

which returns .936 . . . For $p = .05$, just move the cursor up, change the 6 to a 5 throughout, and press ENTER to get .954 . . . This is as close as we can get to $P_a = .95$, so the AQL is 5%. For the LTPD, move the cursor up, change the 05 to 52 throughout, and press ENTER to get .065 . . . A few more tries gives .100 . . . for $p = .47$. This is as close as we can get to $P_a = .10$, so the LTPD is 47%.

155. Sample Size for .10 or Less Chance of Acceptance

38. This is a finite population problem that calls for the hypergeometric distribution. With x for the variable number of defectives in the sample, we have

$$P(x \text{ or less}) = \frac{\sum_{i=0}^{x} \binom{k}{i}\binom{N-k}{n-i}}{\binom{N}{n}}$$

where the $\binom{k}{i} = k!/[i!(k-i)!]$ is the number of combinations of k things taken i at a time. The x must be greater than or equal to the bigger of 0 and $n - (N - k)$, and also less than or equal to the smaller of n and k. Here $N = 120$, $k = 6$, $x = 0$, and we want to find the smallest n such that $P(0) \le .10$. There is apparently no solution other than systematic trial and error. On APL, we can execute)LOAD 42 STAT1, and then HYPER 120 6 20 0 gives $P(0) = .32634$ for $n = 20$. For $n = 30$, just move the cursor up, change 20 to 30 and press ENTER to get $P(0) = .17045$. Continuing in this manner shows that $n = 37$ gives $P(0) = .10333$, and $n = 38$ gives $P(0) = .09586$, so $n = 38$ is the required answer.

156. Total Defectives for .10 or Less Chance of Acceptance

38. If you use systematic trial and error as in Problem 155, you will find that HYPER 120 37 6 0 returns $P(0) = .10333$, and HYPER 120 38 6 0 returns $P(0) = .09586$, so $k = 38$ is the required answer. Note these are the same probabilities we saw in the Problem 155 solution. The HYPER input is the same, too, except n and k have been switched. Switching n and k does not change a hypergeometric probability. You can see this if you expand the hypergeometric probability, so it's in terms of the factorials.

157. Upward Trend?

Yes. Plot the $n = 10$ points with $x =$ day, 1 through 10, and y = yield. We can use the t-test on the slope of the least-squares

line, or the *t*-test on the correlation coefficient *r*. The tests give
exactly the same results. Here $r = .595$, and the *t*-test on *r* gives

$$t = \frac{r\sqrt{n-2}}{\sqrt{1-r^2}} = \frac{.595\sqrt{8}}{\sqrt{1-(.595)^2}} = 2.09$$

With $10 - 2 = 8$ degrees of freedom, $t_{.05} = 1.860$, so the observed
t is significant. Alternatively, the critical coefficient r_c for a given
n can be found by setting $t = t_{.05}$ and $r = r_c$, and then solving
for r_c:

$$r_c = \sqrt{\frac{t_{.05}^2}{n-2+t_{.05}^2}}$$

Here we have $r_c = .549$, so the observed *r* of .595 is significant.
A curve of r_c versus *n* is useful and available from me.

158. Thank God It's Friday?

Yes. Do a correlation analysis with $x =$ day of week starting with
Saturday and $y =$ deaths, so the (x,y) pairs are $(1, 119)$, $(2, 110)$,
. . . , $(7, 84)$. This gives a downward linear trend with a correlation coefficient of $r = -.967$. With $n = 7$ points, the *t*-test on
r gives

$$t = \frac{r\sqrt{n-2}}{\sqrt{1-r^2}} = -8.49$$

The two-tail probability for this *t* with $7 - 2 = 5$ degrees of
freedom is virtually zero, so the downward trend is significant.
Watch out for the weekend! It is interesting that the chi-square
test on the daily frequencies *misses* the difference, because order
is not considered.

159. Random Residuals

Nothing. The *r* for *x* and $y - (a + bx)$ is always exactly zero.
See the inside cover of Freund (7) for the well-known *r* and *b*
equations for *n* points. Note that the *r* and *b* numerators are the

same, and if we replace y with $y - (a + bx)$, this common numerator becomes

$$n\Sigma x(y - a - bx) - \Sigma x\Sigma(y - a - bx)$$

$$n\Sigma xy - na\Sigma x - nb\Sigma x^2 - \Sigma x\Sigma y + na\Sigma x + b(\Sigma x)^2$$

$$n\Sigma xy - [b][n\Sigma x^2 - (\Sigma x)^2] - \Sigma x\Sigma y$$

$$n\Sigma xy - \left[\frac{n\Sigma xy - \Sigma x\Sigma y}{n\Sigma x^2 - (\Sigma x)^2}\right][n\Sigma x^2 - (\Sigma x)^2] - \Sigma x\Sigma y$$

$$n\Sigma xy - n\Sigma xy + \Sigma x\Sigma y - \Sigma x\Sigma y$$

$$0$$

160. Intercept/Slope for Least-Squares Line

0%. For n points, Freund (7) gives

$$a = \frac{\Sigma y\Sigma x^2 - \Sigma x\Sigma xy}{n\Sigma x^2 - (\Sigma x)^2}$$

$$b = \frac{n\Sigma xy - \Sigma x\Sigma y}{n\Sigma x^2 - (\Sigma x)^2}$$

Let y' be a true current value. Then $y = ry'$, where r is some unknown number in the interval $1 \pm .02$. If we replace y with $y' = y/r$, then the true intercept and slope are $a' = a/r$ and $b' = b/r$. Hence, the true ΔW is $-a'/b' = -a/b$, so the measurement error in y has no effect!

161. Constant Failure Rate?

No. Set up a c-chart, with c equal to the number of fails in a 1000-hour interval. For a constant failure rate, c should follow a Poisson distribution, with the mean estimated as $\bar{c}$ = fails/intervals = 16/13 = 1.23. Since the standard deviation for a Poisson distribution is the square root of the mean, three-sigma control limits for c are

$$\bar{c} \pm 3\sqrt{\bar{c}} = 1.23 \pm 3\sqrt{1.23}$$

This gives no lower limit and an upper limit of 4.56. Hence, the 5 *is* significantly high.

162. Mean Time Between Failures

50 hours. See Nelson (12), p. 166. When the fails occur at random, the time between failures follows an exponential distribution with MTBF = time/fails, and the failure rate is λ = fails/time, or 1/MTBF. If the component fails are random, then the system fails are too, with the system failure rate equal to the sum of the component failure rates. Here the tester MTBF is

$$\text{MTBF} = \frac{1}{\lambda} = \cfrac{1}{\cfrac{1}{344} + \cfrac{1}{114} + \cfrac{1}{120}} = 50$$

163. Burn-In

325 hours. For a discussion of the exponential distribution, see Nelson (12). The probability of failing by time t is

$$F(t) = 1 - e^{-t/\theta}$$

Setting this equal to .999 and solving for t gives:

$$t = -\theta \ln .001 = 6.908\theta$$

With θ = 47 hours, the required burn-in time is 325 hours, or about 2 weeks.

164. Tool Fixed?

46 hours. For constant failure rate λ, the number of fails x in time t follows a Poisson distribution with mean λt and

$$P(x) = \frac{e^{-\lambda t}(\lambda t)^x}{x!}$$

If x and t are observed, then the upper $100C$ percent confidence limit for λ is that value of λ for which $P(x$ or less$)$ is $1 - C$. With

$x = 0$ and $C = .90$, this simplifies to

$$P(0) = e^{-\lambda t} = .10$$

$$\lambda = \frac{-\ln .10}{t}$$

Here we want the upper limit to be the historical $1/20 = .05$, so setting $\lambda = .05$ and solving for t gives $t = 46$.

165. 1% Fail Point for Lognormal Distribution

2932. Since $\ln t$ is normally distributed, the lower 1% point for $\ln t$ must be 2.326 standard deviations below the mean, where 2.326 is the standardized normal variable z for 1% in the tail. We have then

$$P(\ln t \leq \mu_{\ln t} - 2.326\sigma_{\ln t}) = .01$$

$$P(t \leq e^{\mu_{\ln t} - 2.326\sigma_{\ln t}}) = .01$$

The expression on the right side of the inequality is the desired $t_{.01}$. Substituting $\mu_{\ln t} = 8.709$ and $\sigma_{\ln t} = .312$ gives $t_{.01} = 2932$.

166. A Bad Connection

$10(1 - fp)$ for No Replug, and the same for Replug. If $P(t)$ is the probability that a module gets t hours of burn-in, then the mean t is

$$\mu_t = \Sigma t P(t)$$

$P(t)$ can be found with the multiplication and addition probability rules found in most statistics books:

	$P(t)$	
t	No Replug	Replug
0	fp	fp^2
5	0	$2fp(1 - p)$
10	$f(1 - p) + (1 - f) = 1 - fp$	$f(1 - p)^2 + (1 - f)$

A little algebra shows that μ_t is $10(1 - fp)$ for either method. Note that for $t = 0$, fp^2 is always less than fp, so Replug gives fewer modules with no burn-in. However, for $t = 10$, it can be shown that $f(1 - p)^2 + (1 - f)$ is always less than $1 - fp$, so Replug also gives fewer modules with the full 10 hours of burn-in. Hence, even though Replug does not increase the mean amount of burn-in, it does spread the burn-in over more modules.

167. Zero Fails

.6557%. We can use a binomial distribution model with n = trials = units, x = occurrences = fails, and p = population fraction failing. The upper $100(1 - \alpha)\%$ confidence limit for p, call it p_U, is that value of p such that the probability of x or less is α. When $x = 0$, we have

$$(1 - p_U)^n = \alpha$$
$$p_U = 1 - \alpha^{\frac{1}{n}}$$

For $\alpha = .10$ and $n = 350$, this gives $p_U = .006557$, or .6557%, or 6557 parts per million (ppm).

168. Zero Fails and Constant Failure Rate

.6557%. For a constant failure rate λ, the time-to-failure t follows an exponential distribution, with the probability of failing by time t given by

$$p = 1 - e^{-\lambda t}$$

In this equation, an upper limit for λ gives an upper limit for p. Let λ_U be the upper $100(1 - \alpha)\%$ confidence limit for λ. An expression for λ_U is given on p. 55 of Tobias and Trindade (16) and can be derived as follows. λ_U is that value of λ for which the probability of x or fewer fails in time t is α. For $x = 0$, this is the probability of *not* failing in time t, or $e^{-\lambda t}$. If n units are each tested for t_0, so $t = nt_0$, then we have

$$e^{-\lambda_U(nt_0)} = \alpha$$

$$\lambda_U = -\frac{\ln \alpha}{nt_0}$$

Substituting this λ_U for λ, and t_0 for t, in the p equation gives

$$p_U = 1 - e^{\frac{\ln \alpha}{n}} = 1 - (e^{\ln \alpha})^{\frac{1}{n}} = 1 - \alpha^{\frac{1}{n}}$$

Usually, an assumption about the population, like the constant failure rate here, effectively increases the sample size, so the confidence interval becomes narrower, but that's not the case here.

169. Zero Fails and Constant Failure Rate (continued)

.3284%. Let kt_0 be the new time for which we want p_U. In the Problem 168 solution, substituting the same λ_U for λ, but kt_0 for t, in the p equation gives

$$p_U = 1 - e^{\frac{k \ln \alpha}{n}} = 1 - (e^{\ln \alpha})^{\frac{k}{n}} = 1 - \alpha^{\frac{k}{n}}$$

For $k = 250/500 = .5$, $\alpha = .10$, and $n = 350$, this gives $p_U = .003284$, or .3284%. Now the constant failure rate assumption *does* make the confidence interval narrower: 0 to .3284%, as opposed to 0 to .6557%. Note that

$$p = 1 - e^{-\lambda t} = 1 - \left[1 - \lambda t + \frac{(\lambda t)^2}{2!} - \frac{(\lambda t)^3}{3!} + \ldots \right]$$

This is approximately λt for small λt, in which case multiplying t by k multiplies p by k, approximately. That's why $.3284 \approx .5(.6557)$.

170. Spare Parts

2. Let x be the variable number of fails in a month, and $P(x$ or less) be the probability of x or less. We want to find x such that $P(x$ or less) $\geq .99$. Since the failure rate is constant, x follows a Poisson distribution with mean $\mu_x = .35$. For a table of Poisson probabilities, see Duncan (5), p. 1008. He uses u' instead of μ_x for the mean x, or expected value of x. Opposite $u' = .35$, we see $P(1$ or less) $= .951$ and $P(2$ or less) $= .994$, so 2 is the required answer. See Chapter 20 of Ref. 5 for a discussion of the

Poisson distribution and how to use a c-chart to check for a constant failure rate.

171. Finding the Total from Moving Sums

3.9%. Let x_i be the defectives for the ith week, and y_i the sum of defectives for the 13 weeks starting with week i. The desired percent defective for the year is

$$100 \frac{x_1 + x_2 + \ldots + x_{52}}{5200} = 100 \frac{y_1 + y_{14} + y_{27} + y_{40}}{5200}$$

Here $y_1 = 51$, $y_{14} = 50$, $y_{27} = 58$, and $y_{40} = 43$, which gives 3.9%.

172. Trend Analysis on Moving Averages or Sums

I think moving averages or sums should not be used for trend analysis, because they will show apparent trends even if the individual points are random. This is called the Slutzky-Yule effect, after the two statisticians who discovered it. See Kenney and Keeping (11) or L. S. Nelson's "The Deceptiveness of Moving Averages," *Journal of Quality Technology*, April 1983, pp. 99–100. Successive moving averages or sums of k values have $k - 1$ values in common, and this positive correlation causes the false trends. Any trend analysis must be done on the individual, in-

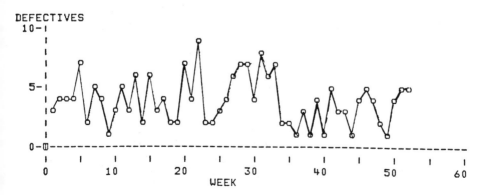

dependent points. Few people would doubt the randomness of the points on the graph. They were, in fact, generated on APL by first executing)COPY 42 STAT13 BINORAND, and then X ← 52 BINORAND 100 .04 to get 52 random x values from the binomial distribution with $n = 100$ and $p = .04$.

173. Finding Individuals from Moving Sums

It can't be done. With x_i and y_i defined as in the Problem 171 solution, we have

$$y_i = x_i + x_{i+1} + \ldots x_{i+12}$$

Since $i = 1, 2, \ldots, 40$, this gives 40 linear equations in 52 unknowns, for which there are many solutions, not just one.

174. 0/n vs. n/n

3. If Method 1 gives x_1/n_1 proportion defective, and Method 2 gives x_2/n_2, then the difference is significant with Fisher's Exact Test if

$$\frac{\sum_{x=0}^{x_1} \binom{n_1}{x}\binom{n_2}{x_1 + x_2 - x}}{\binom{n_1 + n_2}{x_1 + x_2}} \le .05,$$

where $\binom{n}{x} = n!/[x!(n - x)!]$ is the number of combinations of n things taken x at a time. Here $x_1 = 0$, $x_2 = n_2$, and $n_1 = n_2 = n$, so we have

$$\frac{1}{\binom{2n}{n}} = \frac{n!n!}{(2n)!}$$

Trial and error shows that $n = 3$ gives exactly .05, so that's the required sample size. For a derivation of the test and an example, see Duncan (5), Appendix I (31), pp. 973–975.

175. 0/n vs. 1/1

19. With Fisher's Exact Test for two proportions, here $0/n$ versus $1/1$, the chance probability is

$$\frac{\binom{n}{0}\binom{1}{1}}{\binom{n+1}{1}} = \frac{1}{n+1}$$

Note this is one divided by the total number of trials, which could be found directly by reasoning that the one nondefective is equally likely to be in any of the $n + 1$ ordered positions of output. For $1/(n + 1) = .05$, n must be 19.

176. x_1/n_1 vs. $0/n_2$

40. With Fisher's Exact Test for two proportions, we must have

$$\frac{\binom{n_1}{x_1}\binom{n_2}{0}}{\binom{n_1+n_2}{x_1+0}}$$

$$= \frac{n_1(n_1 - 1) \ldots (n_1 - x_1 + 1)}{(n_1 + n_2)(n_1 + n_2 - 1) \ldots (n_1 + n_2 - x_1 + 1)} \leq .05$$

The left-hand side of this inequality is approximately $[n_1/(n_1 + n_2)]^{x_1}$. Set this equal to .05 and solve for n_2 by raising each side of the equation to the $1/x_1$ power. Then use trial and error in the neighborhood of this n_2 to get the exact solution. Here we have $[50/(50 + n_2)]^5 = .05$, which gives $n_2 = 41$. The exact probability for $n_2 = 41$ is

$$(50/91)(49/90)(48/89)(47/88)(46/87) = .046$$

$n_2 = 40$ gives .048, and $n_2 = 39$ gives .051, so $n_2 = 40$ is the answer.

177. x_1/n_1 vs. x_2/n_2

.062. For this type of problem, you could use Fisher's Exact Test for two proportions, but the calculations can be lengthy. A simple alternative with intuitive appeal is the Balls/Cells approximation, which gives nearly the same results if the fraction defective for the pooled samples is less than .20. Here we have $(4 + 0)/(200 + 200) = .01 \ll .20$. Think of the defectives as balls being placed at random into two cells, control and experimental. This is a binomial distribution model with $n = 4$ and $p = .5$. The probability of such an unequal split in the direction observed is $(.5)^4$, or .062, just over the usual .05 required for significance. Fisher's Exact Test gives the same thing, to three decimal places. For more on this, see Problem 126.

178. Old Process Percent Defective vs. New

.18. The 960 units are not a *random* sample from the process because they're restricted to just six lots, which are the *real* random sample. Let x be the sample defectives from a lot, and p be $x/160$. Use the t-test on $\bar{p} = .045$ against the historical $p' = .059$. The standard deviation of $\bar{p}$ is estimated as $s_{\bar{p}} = s_p/\sqrt{n}$, with s_p from the usual standard deviation formula on the $n = 6$ values of p. Those wanting to work with x can use $s_p = s_x/160$. The t is

$$t = \frac{\bar{p} - p'}{s_{\bar{p}}} = \frac{-.014}{.0141} = -.99$$

For $n - 1 = 5$ degrees of freedom, the one-tail probability from 5 TDTR $-.99$ (APL public library 42 STAT2) is .18. By contrast, if we wrongly use the binomial distribution with $n = 960$ and $p' = .059$, the standard deviation of $\bar{p}$ is $\sqrt{.059(.941)/960}$, or .0076, and t is -1.84. The degrees of freedom is infinity because p' is a population value, so t is z, the standardized normal variable. For $z = -1.84$, the z-table gives a one-tail probability of .03, quite different from the correct .18. For a confidence interval approach to this problem, see Problem 192.

179. Fails for Four Tools

No. Use Analysis of Means (ANOM), a modified control chart technique that takes into account the number of points being compared. See Ott (13), pp. 110–111, for an example similar to this problem. If c is the observed fails for a tool, then decision limits are

$$\bar{c} \pm 2.15\sqrt{\bar{c}} = 6.0 \pm 2.15\sqrt{6.0}$$

The 2.15 is a special factor that takes into account that there are just 4 points. For 20 points, the factor is 2.94, approximately the 3 used for process control charts. Here the limits turn out to be .7 to 11.3. Since all the points are within the limits, the differences are not significant.

180. Picking the Better Mean

90%. The manager has been reading Gibbons, Olkin, and Sobel (8), and he used their sample size formula on page 36:

$$n = \left(\frac{\tau\sigma_x}{\delta}\right)^2$$

Here τ is a tabulated factor that depends on the number of populations being compared and also the desired confidence of a correct selection, and δ is the positive difference in population means, $\mu_2 - \mu_1$. For two populations and 90% confidence, $\tau = 1.8124$. With $\delta = 30$ and $\sigma_x = 74$, we have $n = 20$. It is easy to

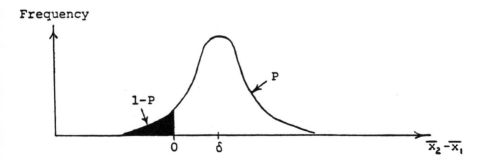

derive the sample size formula for two populations and $100P$ percent confidence of a correct selection. The difference in sample means, $\bar{x}_2 - \bar{x}_1$, follows a normal distribution with mean δ and standard deviation $\sqrt{(\sigma_x^2/n) + (\sigma_x^2/n)}$. If z is the absolute value of the standardized normal variable for $1 - P$ in the tail, then δ must be z standard deviations to the right of zero:

$$\delta = z\sqrt{\frac{\sigma_x^2}{n} + \frac{\sigma_x^2}{n}} \Rightarrow n = \left(\frac{\sqrt{2}z\sigma_x}{\delta}\right)^2$$

Letting $\tau = \sqrt{2}\,z$ gives the formula of Gibbons et al. (8). For $P = .90$, we have $\tau = \sqrt{2}(1.28155) = 1.8124$.

181. Picking the Better Standard Deviation

22. See Gibbons, Olkin, and Sobel (8), Chapter 5 and Table G.1. The table values can be derived as follows. Let σ_1 be the smaller population standard deviation, σ_2 the larger, and s_1 and s_2 the sample standard deviations. We require n such that

$$P\left(\frac{s_1}{s_2} < 1\right) = .90$$

$$P\left(\frac{s_1^2\sigma_2^2}{s_2^2\sigma_1^2} < \frac{\sigma_2^2}{\sigma_1^2}\right) = .90$$

From Duncan (5), Chapter 26, the ratio $s_1^2\sigma_2^2/s_2^2\sigma_1^2$ follows the F distribution with $n - 1$ degrees of freedom for the numerator, and the same for the denominator. With $\Delta = \sigma_1/\sigma_2$, we have, then,

$$P(F < 1/\Delta^2) = .90$$

Here $\Delta = .75$, so $1/\Delta^2 = 1.78$. Systematic trial and error shows $n = 22$ gives the nearest probability to .90. The APL program FDTR is convenient for this. First, execute)COPY 42 STAT2 FDTR NDTR. Then, for example, 21 21 FDTR 1.78 returns .9027 . . . , for $n - 1 = 21$.

182. Picking the Lower Percent Defective

90%. Engineering apparently used the sample size table in my "An Exciting Alternative to Fisher's Exact Test for Two Proportions" (*Journal of Quality Technology*, July 1985, pp. 128–133). Here $p_1 = .01$, $p_2 = .02$, $n = 480$, and $\Pr[X_i = x]$ is given by the binomial distribution. The probability of picking the new method must be

$$P = \sum_{x=0}^{n} (\Pr[X_1 = x])(\Pr[X_2 = x]/2 + \Pr[X_2 \geq x + 1])$$

On APL, P could be found as follows. First, executive)COPY 42 STAT1 BINO BINOCHECK. Then .01 .00005 BINO 480 gives all $\Pr[X_1 = x]$ that are .00005 or more, and .02 .00005 BINO 480 does the same for $\Pr[X_2 = x]$. For $x \geq 16$, $\Pr[X_1 = x]$ is zero to four places, so we only need the probabilities up to $x = 15$. These probabilities can be stored as P1 and P2 by typing P1 ← 80 389 . . . 1 ÷ 10000 and P2 ← 1 6 . . . 277 ÷ 10000. Finally, +/P1 × (.5 × P2) + 1 − +/P2 returns $P = .90$, or 90%.

183. Comparing Machines

No. This data is from L. S. Nelson's "Technical Aids" column in the July 1984 issue of the *Journal of Quality Technology*, pp. 175–176. Nelson's point is that two significant digits are all that's needed to show data like this. Having more than two is really unnecessary and tends to obscure any message the data might have for us. Further insight (not given in Nelson's column) can be achieved by ranking the values within each row, with any ties getting the average of the tied ranks:

Month	Machine A	B	C	D
Jan.	3	4	1	2
Feb.	3	2	4	1
March	3	4	2	1
April	4	2	1	3
May	3	1	2	4
Avg.	3.2	2.6	2.0	2.2

For N rows, k columns, and sum of ranks R_j in the jth column, Friedman's two-way analysis of variance given in Siegel (15) tests the rank averages for significance by calculating

$$\chi_r^2 = \frac{12}{Nk(k+1)} \sum_{j=1}^{k} (R_j)^2 - 3N(k+1)$$

The χ_r^2 is distributed approximately as chi square with $k - 1$ degrees of freedom. Here we have $\chi_r^2 = 2.52$ for an upper tail probability of .47, well over the .05 critical value. We still might suspect problems with machine A, but the test doesn't support this.

184. Fisher's Test?

Yes. Note the A and B samples are *related* and not independent, since the A and B components are on the same 124 units. Hence, we shouldn't use Fisher's Exact Test or any other test for independent samples. The appropriate test is the Sign Test on the 7 units with the A quality worse (+) versus the 1 unit with the A quality better (−). Under the hypothesis of no difference between A and B, we expect just as many + 's as − 's, and the binomial distribution gives the probability of a 7-to-1 split or worse in the observed direction—that is, A worse. With $n = 8$ and $p = .5$, we have $P(1 \text{ or less}) = .035$. Since this is less than .05, the difference is significant. This is a one-tailed test that A is worse.

185. Inspectors/Several Units Inspected Once

Yes. This is an adaptation of a diphtheria media problem given in W. G. Cochran's "The Comparison of Percentages in Matched Samples," *Biometrika*, Vol. 37, 1950, pp. 256–266. In the paper, Cochran gives his Q test for samples that are *related* and not independent. The test is also given in Siegel (15). For sample size N and k samples, imagine the results in a matrix of N rows and k columns with 1 for bad and 0 for good. Let L_i be the total for the ith row, G_j the total for the jth column, and

$$Q = \frac{k(k-1) \sum\limits_{j=1}^{k} (G_j - \overline{G})^2}{k \sum\limits_{i=1}^{N} L_i - \sum\limits_{i=1}^{N} L_i^2}$$

Under the hypothesis of no difference in populations, Q is distributed approximately as chi square with $k - 1$ degrees of freedom. Here we have $Q = 8.05$ for an upper tail probability of .045, under the .05 critical value.

186. Inspectors/One Unit Inspected Several Times

No. The mean defects by tool are 2.0, 1.7, and 2.0. These means can be tested for significance with the Kruskal-Wallis Test, which is a nonparametric test for three or more samples. See Siegel (15). For an APL program, execute)COPY 42 STAT22 KRUSWAL RANKS CDTR, and then N KRUSWAL X, where N is the vector of sample sizes and X is the vector of data: sample 1, sample 2, . . . The "1-chisquare probability" printed out is the chance probability to be compared with .05. For the present problem, the probability is .763, well over the .05 critical value.

187. Before vs. After

Yes. After the change, the mean x/n is .944, less than the .965 before. Hence, the sample quality *is* worse. Note the x distributions are quite skewed to the left, so we shouldn't use the usual t test on these means, since that test assumes a normal distribution for x. A good alternative which makes no distribution assumption is the Mann-Whitney U Test, given in Siegel (15). For an APL program, execute)COPY 42 STAT22 UTEST RANKS NDTR, and then X UTEST Y, where X and Y are the vectors of data for the two samples. The "significance level" printed out is the one-tail probability based on a normal approximation. For the present problem, UTEST gives a probability of .041, less than the .05 critical value. It is interesting that the usual t test gives a probability of .13, which is *not* significant.

188. Suicide Prevention Test

Yes! First, translate the newspaper figures into a table.

	Score		
Suicide	Low	High	All
Yes	1	13	14
No	153	40	193
All	154	53	207

Now we can see that the suicide rate is $1/154 = .6\%$ for the low scorers, and $13/53 = 24.5\%$ for the high scorers. Fisher's Exact Test (Problem 174) gives a one-tail probability of .00000. This can be found with HYPER (Problem 86) with $N = 154 + 53 = 207$, $k = 154$, $n = 1 + 13 = 14$, and $x = 1$: HYPER 207 154 14 1. Some texts would use $k = 14$ and $n = 154$, but interchanging n and k like this does not affect a hypergeometric probability. For a conversational Fisher's Exact Test, execute)COPY 42 STAT3 FISHERΔEXACT PROMPT, and then 1 154 FISHERΔEXACT 13 53.

189. Within-Lot Variability

Yes. This problem is based on Case History 13-1 in Ott (13), pp. 248–251. Ott uses a range chart on the six sample ranges: 2.8, 3.0, 2.4, 1.1, 1.1, and 2.5. With R = range, this leads to $\overline{R} = 2.15$ and three-sigma control limits:

$$\text{UCL} = D_4\overline{R} = 1.92(2.15) = 4.13$$
$$\text{LCL} = D_3\overline{R} = .08(2.15) = .17$$

All the R values are within the limits, so we conclude there is no significant difference in the within-sample variability. For other tests, see Duncan (5), pp. 762–765.

190. Lot Means

No. Use Ott's Analysis of Means (ANOM), a modified control chart technique that takes into account the number of points being compared and the degrees of freedom for the standard deviation estimate. Here the $\overline{X}$ values are: 16.0, 15.8, 16.4, 16.5, 16.1, and 15.0. First, we estimate the within-lot standard deviation as

$$\hat{\sigma}_x = \frac{\overline{R}}{d_2^*} = \frac{2.15}{2.73} = .788$$

The estimate for the standard deviation of $\overline{X}$ with sample size $r = 7$ is

$$\hat{\sigma}_{\overline{x}} = \frac{\hat{\sigma}_x}{\sqrt{r}} = \frac{.788}{\sqrt{7}} = .30$$

For k samples each of size r, the degrees of freedom for $\sigma_{\overline{x}}$ is $df \simeq .9k(r - 1)$, which here is 32. Decision limits for the sample means are then

$$\overline{\overline{X}} \pm H_{.05} \, \hat{\sigma}_{\overline{x}} = 15.97 \pm 2.55(.30) = 15.97 \pm .77$$

This gives UDL $=$ 16.74 and LDL $=$ 15.20. Since the 15.0 is out, we conclude that the vendor's process mean is not stable.

191. Sample vs. Rest of Lot

No. This is a finite population problem that calls for the hypergeometric distribution. Here we have a lot of $N = 250$ modules with a total of $k = 1 + 15 = 16$ defectives. We want to know if it is reasonable to inspect a random sample of $n = 32$ modules and see $x = 1$ or less defectives. In general, the probability of x or less is given by the hypergeometric distribution:

$$P(x \text{ or less}) = \frac{\sum_{i=0}^{x} \binom{k}{i}\binom{N - k}{n - i}}{\binom{N}{n}},$$

where the $\binom{k}{i} = k!/[i!(k - i)!]$ is the number of combinations of k things taken i at a time. The x must be greater than or equal to the bigger of 0 and $n - (N - k)$, and also less than or equal to the smaller of n and k. On APL, we can execute)LOAD 42 STAT1, and then HYPER 250 16 32 1 gives $P(1$ or less) $= .36563$, or about one time in three. This can hardly be considered an unusual event. The .36563 is well over the usual critical value of .05 for a "significant difference," so we conclude that Plant B has no case against Plant A for its inspection. Plant A could do some work on improving quality, though.

192. Confidence Interval for New Process Percent Defective

.9% to 8.1%. We have no right to treat the 960 units as a random sample from a stable process. The $n = 6$ lots are the real random sample here. Let x = sample defectives for a lot, $p = x/160$, $\bar{p}$ = sample mean, and s_p = sample standard deviation. Those wanting to work with x can use $\bar{p} = \bar{x}/160$ and $s_p = s_x/160$. According to the Central Limit Theorem, $\bar{p}$ is approximately normal, even if p is not. Hence, approximate 95% confidence limits for the mean value of p are

$$\bar{p} \pm \frac{ts_p}{\sqrt{n}}$$

The t is Student's t for $n - 1$ degrees of freedom and two-tail area equal to one minus the confidence as a fraction: $1 - .95 = .05$ here. Values for this problem are $\bar{p} = .045$, $t = 2.571$, and $s_p/\sqrt{n} = .0141$. This gives the limits of .9% and 8.1%. By contrast, if we wrongly use a binomial distribution model for a stable process and $n = 960$, the limits would be 3.3% and 6.0%, too narrow an interval for the way this data was taken. On APL, these limits can be found with)LOAD 42 STAT2, and then .95 BICI 43 960 returns the limits as fractions. For a significance test approach to this problem, see Problem 178. For the connection between significance tests and confidence intervals, see the last sentence in Solution 130.

193. Split Lots

3.2% to 7.4%. Let Δ be the difference in yields for a split lot, experimental minus control. Then for n split lots with mean $\bar{\Delta}$

and standard deviation s_Δ, the 95% confidence interval for the population mean is

$$\overline{\Delta} \pm \frac{ts_\Delta}{\sqrt{n}}$$

where t is Student's t with $n - 1$ degrees of freedom and two-tail probability of .05. Here $n = 10$, $\overline{\Delta} = 5.3$, $s_\Delta = 2.9$, and $t = 2.262$. Note it would be incorrect to treat the control and experimental samples as independent, because split lot samples are *not* independent, by design.

194. Making a Good Widget

4. Let p be the fraction defective, and n the number of starts for $100C$ percent confidence that at least one is good. Then the probability that all n widgets are bad is p^n, and the probability of at least one good is $1 - p^n$. Setting this last probability equal to the desired C and solving for n gives:

$$n = \frac{\log (1 - C)}{\log p}$$

For $p = .53$ and $C = .90$, n is 3.6, or 4 to the nearest integer.

195. Winning at Solitaire

23% to 49%. Since the wins and losses are both well over five, the normal distribution approximation to the binomial distribution will do the job here. If $100p$ is the observed winning percent in n trials, then 95% confidence limits for the population winning percent are

$$100p \pm 1.96 \sqrt{\frac{100p(100 - 100p)}{n}}$$

With $p = 18/50 = .36$ and $n = 50$, we have 36 ± 13, or 23% to 49%. The interval is wide, but it does tell us that the game is not fair to the player, since 50% is not in the interval.

196. Total Units in Stock

440,477 to 529,883. See Cochran (1). The estimate and error for the *total* are N times the estimate and error for the *mean*. With t equal to Student's t for 95% confidence and $n - 1$ degrees of freedom, we have.

$$N\bar{x} + Nt\left(\frac{s_x}{\sqrt{n}}\right)\sqrt{1 - \frac{n}{N}}$$

Here we have $t = 2.262$, and the limits are $485,180 \pm 44,703$.

197. Distribution-Free Tolerance Limits

95%. See Duncan (5), pp. 146–147. For confidence 100ε percent that at least 100β percent of the population is included between the lowest and highest values in a sample of size n, the exact relation among the three variables is

$$n\beta^{n-1} - (n - 1)\beta^n = 1 - \varepsilon$$

This is difficult to solve for beta, given n and ε. However, Duncan (5) gives the following approximate relation:

$$n \simeq \frac{\chi^2_{1-\varepsilon,4}(1 + \beta)}{4(1 - \beta)}$$

where $\chi^2_{1-\varepsilon,4}$ is the chi-square value for right-hand probability $1 - \varepsilon$ and 4 degrees of freedom. This equation can be easily solved for β:

$$\beta \simeq \frac{4n - \chi^2_{1-\varepsilon,4}}{4n + \chi^2_{1-\varepsilon,4}}$$

For our problem, Duncan's Table C (5) gives $\chi^2_{.05,4} = 9.49$, and this along with $n = 100$ gives $\beta \simeq .95$, or 95%.

198. Normal-Distribution Tolerance Limits

3.3, 16.3. Dixon and Massey (3) has a table of tolerance factors K for a normal distribution. Here the required limits are

$$\bar{x} + Ks_x = 9.8 \pm (2.934)(2.2) = 9.8 \pm 6.5$$

199. Normal-Distribution Tolerance Limits (*continued*)

4.9, 14.7. Dixon and Massey's table (3) gives $K = 2.233$, so the limits are

$$\bar{x} \pm Ks_x = 9.8 \pm 2.233(2.2) = 9.8 \pm 4.9$$

These limits are different from those for Problem 197, but not a contradiction. There is more than one way to include at least 95% of a population. Note that the Problem 197 limits, with no distribution assumption, are 10% wider:

$$\frac{15.0 - 4.2}{14.7 - 4.9} = \frac{10.8}{9.8} = 1.10$$

200. Distribution-Free Confidence Limits for Percent Defective

2.0%, 5.3%. The estimate is the $2/100 = 2.0\%$ observed in the sample. The upper limit is the percent defective for which the probability of 2 or less in 100 is: 100% minus the 90% confidence, or 10%. This is exactly the same as the lot tolerance percent defective (LTPD) for the acceptance sampling plan with sample size $n = 100$ and acceptance number $c = 2$. This can be found in several ways, but perhaps the simplest is F. E. Grubbs' table in Duncan (5), p. 172. This is based on the Poisson distribution, a good approximation to the binomial distribution if the population fraction defective p is less than .10. For $c = 2$ and 10% chance of acceptance, Grubbs' table gives $np = 100p = 5.32$, so $p = .0532$ or 5.3%.

201. Normal-Distribution Confidence Limits for Percent Defective

4.2%, 6.3%. For the estimate, calculate the standardized normal variable $z = (L - \bar{x})/s_x = -1.73$ and get 4.2% from the z table. As in the previous problem, the upper limit is the LTPD for the acceptance sampling plan with the acceptance limit right at the observed sample value. See Duncan (5), p. 277. To meet the p'_2, β criteria, we must have

$$\frac{-z_2 + k}{\sqrt{\dfrac{1}{n} + \dfrac{k^2}{2n}}} = z_\beta$$

In this formula, $100p_2'$ is the LTPD, β is the chance of acceptance, z_2 and z_β are the standardized normal variables for right-hand tail areas of p_2' and β, n is the sample size, and k is the acceptance limit. We accept if $(\bar{x} - L)/s_x \geq k$. Here $\beta = .10$, $z_\beta = 1.282$, and $n = 100$. Use the observed $(\bar{x} - L)/s_x = 1.73$ for k, and solve for z_2:

$$z_2 = k - \frac{z_\beta}{\sqrt{n}} \sqrt{1 + \frac{k^2}{2}} = 1.53$$

The z table gives $100\,p_2' = 6.3\%$. Note that the Problem 200 approach, with no distribution assumption, gives poorer precision: $5.3 - 2.0 = 3.3$ as opposed to $6.3 - 4.2 = 2.1$.

202. Election Night

Yes. We want to know the confidence that Dean will get more than 50% of the vote, so Dean is a clear-cut winner, and the legislature stays out of it. See Duncan (5), pp. 572–575. Let N = population size, n = sample size, p = sample proportion, and p' = population proportion. Then the estimate for the standard deviation of p is

$$\hat{\sigma}_p = \sqrt{\frac{p(1 - p)}{n}} \sqrt{1 - \frac{n}{N}}$$

The lower 100C% confidence limit for p' is

$$p_L' = p - z\hat{\sigma}_p$$

where z is the standardized normal variable for $100C\%$ cumulative probability. Setting $pL = .50$ and solving for z gives

$$z = \frac{p - .50}{\hat{\sigma}_p}$$

Here $n = 55{,}731$, $n/N = .41$, and $p = 29{,}264/55{,}731$, or $.5251$, for Dean. This leads to $z = 15.4$, for which $100C\%$ is virtually 100%.

203. Two-Stage Sample for Percent Defective

12.4% to 83.6%. See Cochran (1), one of the few books that deals with samples other than simple random samples. Here we have a two-stage sample: a random sample of primary units (wafers), and then a random sample of secondary units (chips) from each of the selected primary units. With p_i equal to the sample fraction defective for the ith wafer, and $\bar{p} = \Sigma p_i/n$, let

$$s_1^2 = \frac{\Sigma(p_i - \bar{p})^2}{n - 1} \qquad s_2^2 = \frac{m}{n(m - 1)} \Sigma p_i(1 - p_i)$$

Then an unbiased estimate of the variance of $\bar{p}$, with $f_1 = n/N$ and $f_2 = m/M$, is

$$v(\bar{p}) = \frac{1 - f_1}{n} s_1^2 + \frac{f_1(1 - f_2)}{mn} s_2^2$$

According to the Central Limit Theorem, $\bar{p}$ is approximately normal, so approximate 95% confidence limits for the lot fraction defective are

$$\bar{p} \pm t\sqrt{v(\bar{p})}$$

where t is Student's t for $n - 1$ degrees of freedom and two-tail area equal to one minus the confidence as a fraction: $1 - .95 = .05$ here. Values for this problem are $\bar{p} = .480$, $t = 2.776$, and $v(\bar{p}) = .01624 + .00024 = .01648$. This gives $48.0\% \pm 35.6\%$, or 12.4% to 83.6%. By contrast, if we incorrectly treat the sample as a simple random sample and use the usual normal approximation, we have

$$\bar{p} \pm 1.96 \sqrt{\frac{\bar{p}(1 - \bar{p})}{mn}} \sqrt{1 - \frac{mn}{MN}}$$

This gives 48.0% ± 13.8%, too narrow an interval for the way this data was taken.

204. Tool Set-Up

10. See Duncan (5), pp. 563–567. The maximum sampling error in $\bar{x}$, with 95% confidence, is

$$E = 1.96\sigma_{\bar{x}} = 1.96 \frac{\sigma_x}{\sqrt{n}} \Rightarrow n = \left(\frac{1.96\sigma_x}{E}\right)^2$$

$E = .5$ and $\sigma_x = .82$ gives $n = 10.3$, or 10 to the nearest integer.

205. Repeatability Error

1.98. See Duncan (5), pp. 568–569. Let the chi-square value with right-hand probability P and degrees of freedom $v = n - 1$ be $\chi^2_{P,n-1}$. Then with 95% confidence, we have

$$\sigma_x^2 \leq \frac{(n - 1)s_x^2}{\chi^2_{.95,n-1}}$$

$$3\sigma_x \leq 3s_x \sqrt{\frac{n - 1}{\chi^2_{.95,n-1}}}$$

The right-hand side of this last inequality is the required upper limit. Here we have $n - 1 = 19$, $\chi^2_{.95,19} = 10.1$, and $3s_x = 1.44$, so the limit is $1.44\sqrt{19/10.1} = 1.44(1.372) = 1.98$. The 1.372 multiplier means that the upper limit is 37.2% higher than the estimate whenever $n = 20$.

206. Repeatability Error (*continued*)

167. We require

$$\frac{3s_x\sqrt{\dfrac{n-1}{\chi^2_{.95,n-1}}} - 3s_x}{3s_x} = .10$$

$$\frac{n-1}{\chi^2_{.95,n-1}} = 1.21$$

Duncan's table (5) only goes to $v = n - 1 = 30$, which isn't far enough for this problem. However, we can use systematic trial and error in APL. Execute)LOAD 42 STAT2 and then 166 ÷ (166 CHISQUARE .05) to get 1.209 . . . , for example. The 166 is the $n - 1$ degrees of freedom, and .05 is the left-hand probability, so (166 CHISQUARE .05) is our $\chi^2_{.95,166}$. This is as close as we can get to 1.21, so the required n is 166 + 1, or 167.

207. Defects in Windows

$.925 \pm .433 = .492$ to 1.358. First of all, $(\Sigma x)/(nA) = \bar{x}/A$, where $\bar{x} = (\Sigma x)/n$ is the sample mean. Here $\bar{x} = 1.850$, so the estimated defect density is $1.850/2$, or .925 defects per square centimeter. The Central Limit Theorem tells us that the distribution of $\bar{x}$ is approximately normal, even if x is not. This is a finite population problem because the unit has $N = T/A = 100/2 = 50$ total windows. Cochran (1) gives the possible plus-or-minus sampling error in $\bar{x}$ from a finite population, which leads to the maximum sampling error in $\bar{x}/A$, with 95% confidence:

$$E_{\frac{\bar{x}}{A}} = \frac{E_{\bar{x}}}{A} = \frac{2\dfrac{s_x}{\sqrt{n}}\sqrt{1 - \dfrac{n}{N}}}{A},$$

where the 2 is an approximation to Student's t for right-hand probability .025 and $n - 1$ degrees of freedom, and s_x is the sample standard deviation. For $n = 20$, $t = 2.093$, and as n increases, t approaches 1.96 as a limit. Hence, for $n \geq 20$, $t = 2$ is a good approximation. The $\sqrt{1 - (n/N)}$ is called the finite population correction factor. Finally, $s_x = 2.498$ here, which gives $E_{\bar{x}/A} = .433$.

208. Defects in Windows (*continued*)

No. Let y be the new defect count for a window. If λ is the true defect density for the unit, then the expected value of Σy is $\lambda(nA)$, the same as the expected value of Σx. Hence, the expected value of the defect density estimate is the same: $(\lambda nA)/(nA) = \lambda$. With the change in window size, the new N is the total area T divided by the new window size kA: $T/(kA) = (T/A)/k = N/K$. Hence, the finite population correction factor doesn't change, because $(n/k)/(N/k) = n/N$. The only thing we don't know is s_y. Since y is k times as big as x *on the average*, it's tempting to say that our best guess for s_y is ks_x. This would be true if we had the exact relation $y = kx$, but that isn't the case. As an example to show the fallacy in this reasoning, suppose the defects are randomly spread across the unit, so x and y each follow a Poisson distribution, with the standard deviation equal to the square root of the mean: $\sigma_x = \sqrt{\lambda A}$ and $\sigma_y = \sqrt{\lambda kA} = \sqrt{k}\,\sigma_x \neq k\sigma_x$. Substitution of these values in the Problem 207 error expression shows that the error is the same regardless of window size. Our data does not follow a Poisson distribution though, so this doesn't help us. For a nonrandom distribution of defects, it does seem that more windows would cover the unit better, so the error should be less, but we can't quantify it. Readers are invited to offer a solution to this problem.

209. Voids in Lines

47.4. Let x be the variable number of voids in length L, and $P(x$ or more) be the probability of x or more. Since the void density is constant, x follows a Poisson distribution with mean $\mu_x = \lambda L$.

Solving for L gives

$$L = \frac{\mu_x}{\lambda}$$

In this equation, we will substitute $\lambda = .3$ and the μ_x such that $P(10$ or more$) = .90$, or equivalently, $P(9$ or less$) = .10$. For a table of Poisson probabilities, see Duncan (5), p. 1008. He uses u' instead of μ_x for the mean x, or expected value of x. $u' = 14.0$ gives $P(9$ or less$) = .109$, and $u' = 14.5$ gives $P(9$ or less$) = .088$. For $P(9$ or less$) = .100$, linear interpolation gives $u' = 14.214$. The required L is $14.214/.3$, or 47.4. See Chapter 20 in Duncan (5) for a discussion of the Poisson distribution, and how to use a c-chart to check for a constant defect density.

210. Voids in Lines (continued)

.91% to 1.35%. For the ith line in the population of N lines, let V_i be the area voided, T the total area, and $y_i = 100(V_i/T)$ be the percent area voided. Then for the population, the percent area voided is

$$100 = \frac{\sum\limits_{i=1}^{N} V_i}{NT} = \frac{\sum\limits_{i=1}^{N} 100\,\dfrac{V_i}{T}}{N} = \frac{\sum\limits_{i=1}^{N} y_i}{N} = \mu_y$$

The problem, then, is just to find 95% confidence limits for the population mean μ_y. The limits are given in Duncan and elsewhere as

$$\bar{y} \pm t_{.05,n-1}\,\frac{s_y}{\sqrt{n}}$$

where $\bar{y}$ is the sample mean, s_y is the sample standard deviation, and $t_{.05,n-1}$ is a tabulated value for Student's t with two-tail area $1 - .95 = .05$ and degrees of freedom $n - 1$. Here we have $n = 5$, $\bar{y} = 1.13$, $t_{.05,n-1} = 2.776$, and $s_y = .1747$, which gives $1.13 \pm .22$, or .91% and 1.35%.

211. Three Navaho Women

The squaw on the hippopotamus is equal to the sons of the squaws on the other two hides.

212. Switched Labels

Take a piece of fruit from the "apples and oranges" crate. That crate gets the label for whatever fruit you pick, "apples" or "oranges," because that crate can't be a mixture. Now the two remaining crates must get different labels. The only way to do this is to have the unlabeled crate get the label from the other one, which gets "apples and oranges."

213. Triangle Former

1/4. A triangle can be formed as long as this does not happen: ⟵————⟶. This will *not* happen if the length of the longest segment—call it d—is less than the sum of the other two. For convenience, pick the points on the number line between zero and one. Then the required condition for successful triangle formation is $d < 1 - d$, or $d < .5$. This means that each of the segment lengths must be less than .5. Suppose x is the first point, and it's between zero and .5. Then the second point must be on the *other* half of the line, and between .5 and .5 + x, as well:

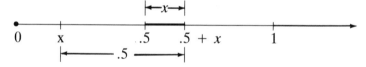

The joint probability that x is in an interval dx between zero and .5, and then the second point is in the shaded, favorable x interval is $(dx/1) \, (x/1)$. Integrating gives

$$\int_0^{.5} x \, dx = \left. \frac{x^2}{2} \right]_0^{.5} = \frac{1}{8}$$

The probability must also be 1/8 for the first point on the right half of the line segment, and then the second point in the favorable interval on the left half. Summing the probabilities for the two mutually exclusive events gives the required answer: 1/8 + 1/8 = 1/4.

214. Potato Race

25 miles. The distances to the piles in yards are found by summing the odd integers, which gives the perfect squares: 1, 4, 9, 16, etc. The boys must go to each pile and return, so the total distance in miles is

$$\frac{2 \sum_{x=1}^{40} x^2}{1760} = \frac{\dfrac{[40(40 + 1)] [2(40) + 1]}{6}}{880} = 25$$

A nice bit of exercise before ice cream and cake.

215. Tester Availability

90%. For n tools and a long clock time T, the scheduled test time is nT. Suppose each tool is down p fraction of the time. Then the actual test time for a single tool is $(1 - p)T$, and for n tools, it's $n(1 - p)T$. It follows that the tester availability is $100n(1 - p)T/nT$, or $100(1 - p)$. Hence, the tester availability is the same as that for each of its tools. For $p = .10$, we have 90%.

216. Math Test

Children have shorter feet and lower math ability than adults.

217. Crossing a Moat

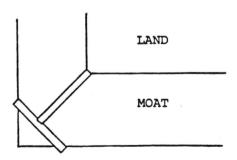

218. Selling Sheep

$2. If N is the number of sheep, then they received N^2 dollars in all. Consult a table of squares, or make your own on APL with $(\iota N)*2$. Note that whenever the 10s position is odd, the units position is 6: 16, 36, 196, etc. This means the number of $1 bills is 6. With x for the amount of the check to make the split equal, we must have: $10 - x = 6 + x$, or $x = 2$.

219. A Boy, a Girl, and a Dog

A half-mile. The boy and girl each walk 1/8 of a mile before they meet. Since time = distance/rate, and the rate is 3 miles per hour, this takes (1/8)/3, or 1/24 of an hour. During this time the dog is running at 12 miles per hour, so distance = (time) (rate) = (1/24) (12) = 1/2 mile. In general, if the initial distance between the boy and the girl is d, then the dog runs $2d$.

220. Bookkeeper

Bookkeeper.

221. Two Coins

A quarter and a nickel. The quarter is not a nickel.

222. Shoeing a Horse

$43 million! In pennies, the total cost is

$$1 + 2 + 4 + \ldots + 2^{31}$$

This is the sum of the first 32 terms of a geometric progression with ratio 2, which reduces to $(1 - 2^{32})/(1-2)$, or 4.3×10^9. Multiplying this by .01 to convert to dollars gives 4.3×10^7, or $43 million.

223. Foot Race

The boy wins. When the boy has run 100 yards, the girl has run 95 yards, so they are dead-even 5 yards from the tape. The boy's superior speed enables him to just win. If the boy lets the girl start 5 yards ahead of the starting line, it *will* be a tie.

224. Average Speed

∞. Let x be the unknown rate. Since time = distance/rate, it takes 1/15 hours to go up and $1/x$ hours to go down. For 30 miles per hour overall, these times must total 2/30, so we have

$$\frac{1}{15} + \frac{1}{x} = \frac{2}{30}$$

But this gives $x = \infty$. Two miles at 30 miles per hour takes 2/30 = 1/15 hours, which is how long it takes just to get to the top of the mountain. The answer isn't 45, even though $(15 + 45)/2 = 30$. That isn't how you figure the overall rate, which is distance/time.

225. Poles With Guidewires

Anything! For a general solution, let $H_1 = 10$, $H_2 = 15$, $y = 6$, and split x into two parts: x_1 to the left of the point on the ground directly below the guidewires' intersection, and x_2 to the right. By similar triangles, we have

$$\frac{y}{H_1} = \frac{x_2}{x_1 + x_2} \text{ and } \frac{y}{H_2} = \frac{x_1}{x_1 + x_2}$$

Adding these two equations and simplifying gives

$$y = \frac{H_1 H_2}{H_1 + H_2}$$

In other words, for a given set of poles, y is always the same, regardless of x.

226. Calculator Trick

4. Try writing down the results for a few steps, and you will see that each result is of the form $2^a x^b$. For steps $1, 2, 3, \ldots, n$, we have

$$a = 1, \frac{3}{2}, \frac{7}{4}, \ldots, \frac{2^n - 1}{2^{n-1}}$$

$$b = \frac{1}{2}, \left(\frac{1}{2}\right)^2, \left(\frac{1}{2}\right)^3, \ldots, \left(\frac{1}{2}\right)^n$$

As n goes to infinity, a goes to 2, and b goes to 0. Hence, the answer is $2^2 x^0 = 4$ for *any* positive, non-zero x. This is fun to do on your calculator, although you may not get exactly 4 due to roundoff error.

227. Another Calculator Trick

No limit. The sixth term equals the first, and the seventh equals the second. Hence, as we extend the sequence indefinitely, the same five numbers repeat themselves over and over.

228. Half-Full

He tilts the glass until the beer is right at the top edge and notes that part of the bottom is not covered with beer. If the bottom were all covered, the glass would be half-full or more.

229. Dog, Goose, and Corn

Use trial and error with three different coins. Let the dog be the biggest coin, the goose the next biggest, and the corn the smallest. One solution is:

Trip	Over	Back
1	Goose	—
2	Dog	Goose
3	Corn	—
4	Goose	—

For the only other solution, interchange dog and corn in the table.

230. Missionaries and Cannibals

Use trial and error with, say, nickels for missionaries and pennies for cannibals. In the following solution, M = missionary, and C = cannibal:

Trip	Over	Back
1	CC	C
2	CC	C
3	MM	MC
4	MM	C
5	CC	C
6	CC	—

For the only other solution, trip 1 could be MC over and M back, and then the following trips are the same as in the table.

231. Picking Grapes

100 yards. One way to do this problem would be to set up an equation for the walking distance, $y = FG + GH$, set $dy/dx = 0$, and solve the quadratic equation for x. Another way is to find an easier problem with the same solution. To do this, continue the perpendicular from F to the grapevine for another 100 yards

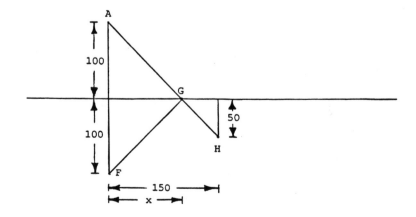

to the point A, and suppose you were asked the shortest distance between A and H. You would certainly say the straight line AH. In other words, the point G as shown minimizes AG + GH. Now note that regardless of where G is along the vine, AG = FG by congruent triangles. Hence, the point G as shown minimizes FG + GH, too. Finally, by similar triangles, we have:

$$x/(150 - x) = 100/50 \Rightarrow x = 100$$

232. Spaghetti Measure

$.875\sqrt{i}$. Let D_i be the diameter for i people, and L be the spaghetti length. The spaghetti forms a cylinder with volume:

$$V_i = \pi\left(\frac{D_i}{2}\right)^2 L \qquad .$$

The ratio V_i/V_1 is $(D_i/D_1)^2$. Setting this equal to the desired ratio i and solving for D_i gives $D_1\sqrt{i}$. If you incorrectly let $D_i = iD_1$, then you would make i times as much spaghetti as needed!

233. Intersecting Circles

69π, or 217. Let A be the overlapping area. Since the area of a circle is π times the square of the radius, the required difference is

$$(169\pi - A) - (100\pi - A) = 69\pi$$

The right angle was a red herring. Note that for *any* two overlapping figures, not just circles, the difference in the nonoverlapping areas is the difference in the total areas.

234. Man in a Coffin

Frank is the father of the man in the coffin.

235. A Circular Swimming Pool

90°. Her path is the triangle ABC:

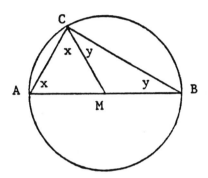

Point M is the middle of the pool. This makes each of the smaller triangles an isosceles triangle, so the angles opposite the equal sides are equal. Regardless of the diameter of the pool or angle y, the angles in triangle ABC (or any triangle) sum to 180°, so

$$x + y + (x + y) = 180$$

$$x + y = 90$$

This proves that an angle inscribed in a semicircle is a right angle.

236. Frog in a Well

38. After 37 days, he has climbed 37 feet. On the 38th day, his 3-foot jump takes him to the top. Free at last!

237. Water in Buckets

Let's work backward. If you could get 8 gallons in the 10-gallon bucket, and then dump 3 into the empty 3-gallon bucket, you would have the 5 you want. To get the 8 in the 10-gallon bucket, you could get 1 there and then add the full 7-gallon bucket to it. To get the 1 in the 10-gallon bucket, you could take the full 10-

gallon bucket and dump it three times into the empty 3-gallon bucket. This can all be done in 9 steps. The buckets' contents after each step are:

Bucket	Step								
	1	2	3	4	5	6	7	8	9
10-gallon	7	7	4	4	1	1	8	8	5
7-gallon	0	3	3	6	6	7	0	2	2
3-gallon	3	0	3	0	3	2	2	0	3

238. Water in Buckets (*continued*)

The buckets' contents after each step are:

Bucket	Step							
	1	2	3	4	5	6	7	8
10-gallon	3	3	6	6	9	9	2	2
7-gallon	7	4	4	1	1	0	7	5
3-gallon	0	3	0	3	0	1	1	3

239. Doubling Your Money

69.7, 72.1, 74.4. If P doubles in n years, we must have $2P = P(1 + i)^n$, so that

$$n = \frac{\ln 2}{\ln(1 + i)} = \frac{D}{100i}$$

$$D = \frac{(100i)(\ln 2)}{\ln(1 + i)}$$

Letting $100i$ be 1, 8, and 15 gives the required answers, which are quite close to one another. Economists have settled on $D \simeq 72$, and their "Rule of 72" is $n \simeq 72/(100i)$.

240. Transportation Problem

You drive the car with the bike on the rack to the repair shop, take the car or bike back home, drive the motorcycle to the shop, and take the car or bike (whichever was left there) back home. After the motorcycle repair work is done, you take the car or bike to the shop, drive the motorcycle home, take the car or bike (whichever was left at home) to the shop, and drive the car with the bike on the rack back home.

241. Summing Integers

5050. Gauss saw that the required sum can be thought of as

$$(1 + 100) + (2 + 99) + (3 + 98) + \ldots$$
$$+ (50 + 51) = 50(101) = 5050$$

In general, for n equal to an even integer, the sum of the integers from 1 to n can be separated into $n/2$ subsums of $n + 1$. If n is odd, then we have $(n - 1)/2$ subsums of $n + 1$, plus the integer midway between 1 and n, which is $(n + 1)/2$. All together, the number of subsums of $n + 1$ is

$$\frac{n - 1}{2} + \frac{1}{2} = \frac{n}{2}$$

Hence, we have the same formula whether n is even or odd: $(n/2)(n + 1)$.

242. Summing Squares of Integers

$n(n + 1)(2n + 1)/6$. The Gauss-like trick is a quite a bit longer:

$$x^3 - (x - 1)^3 = 3x^2 - 3x + 1$$
$$\Sigma[x^3 - (x - 1)^3] = \Sigma[3x^2 - 3x + 1]$$
$$(1^3 - 0^3) + (2^3 - 1^3) + (3^3 - 2^3) + \ldots + [n^3 - (n - 1)^3]$$
$$= 3\Sigma x^2 - 3\Sigma x + n$$
$$n^3 = 3\Sigma x^2 - 3[(n/2)(n + 1)] + n$$

Solving for Σx^2 and simplifying gives

$$\Sigma x^2 = n(n + 1)\,(2n + 1)/6$$

243. Summing Cubes of Integers

4,326,400 for each offer. She must be a very pretty lady! The sums can be found with APL: $(+/\iota 64)*2$ for suitor A, and $+/(\iota 64)*3$ for suitor B. Apparently, $(\Sigma x)^2 = \Sigma x^3$ for $x = 1, 2, 3, \ldots, n$. Problems 241 and 242 were warmups for this one. In Problem 241, we saw that $\Sigma x = (n/2)(n + 1)$. Squaring this gives

$$(\Sigma x)^2 = \frac{n^4 + 2n^3 + n^2}{4}$$

For Σx^3, we follow the method of the Problem 242 solution for Σx^2:

$$x^4 - (x - 1)^4 = 4x^3 - 6x^2 + 4x - 1$$
$$\Sigma[x^4 - (x - 1)^4] = \Sigma[4x^3 - 6x^2 + 4x - 1]$$
$$(1^4 - 0^4) + (2^4 - 1^4) + (3^4 - 2^4) + \ldots + [n^4 - (n - 1)^4]$$
$$= 4\Sigma x^3 - 6\Sigma x^2 + 4\Sigma x - n$$
$$n^4 = 4\Sigma x^3 - 6[n(n + 1)\,(2n + 1)/6] + 4[(n/2)\,(n + 1)] - n$$

Solving for Σx^3 and simplifying gives

$$\Sigma x^3 = \frac{n^4 + 2n^3 + n^2}{4} = (\Sigma x)^2$$

244. Speed of a Pitch

With the stopwatch, see how long it takes for the ball to get to the plate after it leaves the pitcher's hand. If the pitcher strides 4', and his arm goes another 3' closer to the plate, then the distance covered is 53'6". Rate = distance/time gives the answer. I tried this with Roger Clemens of the Boston Red Sox during a

TV game. On one pitch, the announcer said, "That one was 97 miles per hour," and my watch said .38 seconds. Converting the distance to miles, and time to hours, gives

$$\frac{\dfrac{53.5}{5280}}{\dfrac{.38}{3600}} = \frac{36.477}{.38} = 96 \text{ miles per hour}$$

245. Send More Money

SEND = 9567, MORE = 1085, MONEY = 10652. This kind of problem and the following method of solution was suggested by Fred Stoehr, IBM Rochester. Let $w, x, y,$ and z be the carries to the columns:

$$
\begin{array}{cccc}
w & x & y & z \\
S & E & N & D \\
 & M & O & R & E \\
\hline
M & O & N & E & Y
\end{array}
$$

Then we know $w = M = 1$ because the maximum carry is 1. The rest is not so easy:

a. $x + S + 1 = 10 + O$. S has to be 8 or 9 in order to have the $w = 1$ carry. If $S = 8$, then $x = 1$ and $O = 0$. If $S = 9$, then $x = 0$ and $O = 0$. Hence, $O = 0$.

b. $y = 0$ or 1, $O = 0$, and $N \neq 0$ or 1 $\Rightarrow x = 0$, $S = 9$.

c. $y + E + O = y + E = N \Rightarrow y \neq 0$, so $y = 1$, $N = E + 1$.

d. $z + (E + 1) + R = E + 10 \Rightarrow z + 1 + R = 10$. $R \neq 9 \Rightarrow R = 8$, $z = 1$.

e. $D + E$ must be a two-digit number greater than 11. The only possibilities are $7 + 6$ and $7 + 5$, so D or E must be 7. If $E = 7$, then $y + E + O = 8 = N$, which is not right, so $D = 7$.

f. If $E = 6$, then $y + E + O = 7 = N$, which is not right, so $E = 5$, $Y = 2$, and $N = 6$.

246. Triangle Area

1.5. Enclose the triangle in a square:

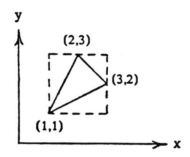

The area of the triangle is the area of the square minus the areas of the three small right triangles:

$$2(2) - \frac{1}{2}(2)(1) - \frac{1}{2}(1)(1) - \frac{1}{2}(1)(2) = 1.5$$

247. Average Time to Get a Good Unit

5 hours. Let N be the long-run number of tries to make a good unit, so that N is the sum of the first tries *and* the reworks. If T is the time to process a unit, and the yield as a fraction is Y, then the average time to get a good unit is

$$\frac{\text{Total time}}{\text{Total good units}} = \frac{NT}{NY} = \frac{T}{Y}$$

The required answer is 2/.4, or 5 hours.

248. Rope Around the Earth

.5 feet. If r_1 is the radius of the earth, r_2 is the radius of the rope-circle, then the rope is $r_2 - r_1$ off the ground. The three feet added to the rope is the difference in the circumferences of the two circles. so we have

$$3 = 2\pi r_2 - 2\pi r_1 = 2\pi(r_2 - r_1) \Rightarrow r_2 - r_1 = \frac{3}{2\pi} = .5$$

Note that we didn't have to use the Earth's diameter of 7913 miles. In fact, the answer to the problem is the same .5 feet for a basketball or any other sphere.

249. Rope Around the Earth (*continued*)

375 feet. In the following figure, the radius of the Earth in feet is $r = .5(7913)(5280)$, and ℓ is the length of the rope from the point of tangency to the highest point:

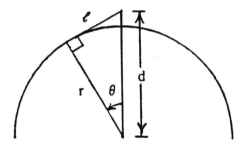

The required height is $d - r$, and since $\cos \theta = r/d$, we have $d - r = (r/\cos \theta) - r$. The hard part is to find θ. If $x = 3$ feet is the length added the rope, then ℓ must be the arc length $r\theta$ plus .5x. This leads to

$$\tan \theta = \frac{\ell}{r} = \frac{r\theta + .5x}{r} = \theta + \frac{.5x}{r}$$

Unfortunately, there appears to be no way to find θ other than systematic trial and error. This can be done on APL or any other computing system that gives $\tan \theta$. With the method of bisection for finding the root of an equation, we let

$$y = \tan \theta - \theta - \frac{.5x}{r} = 0$$

Then we find a θ_1 such that $y_1 < 0$, and a θ_2 such that $y_2 > 0$. $\theta_1 = 0$ and $\theta_2 = \pi/4$ do the job. The desired θ, which gives $y = 0$, must lie between θ_1 and θ_2. We close in on the value by *bisecting* the $[\theta_1, \theta_2]$ interval to find $\bar{\theta} = (\theta_1 + \theta_2)/2$, which gives y_b. If $y_b < 0$, then $\bar{\theta}$ becomes the new θ_1. If $y_b \geq 0$, then $\bar{\theta}$ becomes the new θ_2. This procedure is repeated until $\theta_2 - \theta_1$ is less than some preset number δ. The $\bar{\theta}$ for this interval is within $\delta/2$ of the desired root. Alternatively, we can continue until θ_1 and θ_2 give the same $d - r$ to the nearest foot. This avoids having to figure out how small δ should be. For the 19th interval, $\theta_1 = .005992$ . . . gives $d - r = 375.04$ . . . , and $\theta_2 = .005995$ . . . gives $d - r = 375.41$ . . . Whatever the exact solution is, it's 375 to the nearest foot. You will need a ladder.

250. Buying New Socks

8. Suppose you do your wash on Saturday. Then you need socks in the machine for Sunday through next Saturday, inclusive. That's 7 pairs, and the ones you have on make 8.

251. Chip Height

$z_1 + z_3 - z_2$. Let the xy plane be the substrate, and imagine the chip above it at a variable height z. A point (x, y, z) on the chip plane satisfies the equation $z = ax + by + c$. Position the chip so that the corner with unknown height is at $(0, 0, z_4)$, and the other corners are $(x_1, 0, z_1)$, (x_1, y_1, z_2), and $(0, y_1, z_3)$. Then we have

$$z_1 = ax_1 + c$$
$$z_2 = ax_1 + by_1 + c$$
$$z_3 = by_1 + c$$
$$z_4 = c$$

The problem then is to find c. The third equation gives $c = z_3 - by_1$, and the second equation minus the first gives $z_2 - z_1 = by_1$. Therefore we have

$$z_4 = c = z_3 - by_1 = z_3 - (z_2 - z_1) = z_1 + z_3 - z_2$$

REFERENCES

Books

1. Cochran, W. G., *Sampling Techniques*, 3rd ed., New York: John Wiley and Sons, 1977.
2. Crow, E. L., Davis, F. A., and Maxfield, M. W., *Statistics Manual*, New York: Dover Publications, 1960.
3. Dixon, W. J., and Massey, F. J., Jr., *Introduction to Statistical Analysis*, 4th ed., New York: McGraw-Hill Book Co., 1983.
4. Dodge, H. F., and Romig, H. G., *Sampling Inspection Tables*, 2nd ed., New York: John Wiley and Sons, 1959.
5. Duncan, A. J., *Quality Control and Industrial Statistics*, 5th ed., Homewood, Illinois: Richard D. Irwin, Inc., 1986.
6. Feller, W., *An Introduction to Probability Theory and Its Applications*, Vol. 1, 3rd ed., New York: John Wiley and Sons, 1968.
7. Freund, J. E., *Modern Elementary Statistics*, 7th ed., Englewood Cliffs, N.J.: Prentice-Hill, 1988.

8. Gibbons, J. D., Olkin, I., and Sobel, M., *Selecting and Ordering Populations: A New Statistical Methodology*, New York: John Wiley and Sons, 1977.

9. Grant, E. L., and Leavenworth, R. S., *Statistical Quality Control*, 5th ed., New York: McGraw-Hill Book Co., 1980.

10. Hogg, R. V., and Craig, A. T., *Introduction to Mathematical Statistics*, Macmillan Publishing Co., 1978.

11. Kenney, J. F., and Keeping, E. S., *Mathematics of Statistics*, 3rd ed., Princeton, N.J.: D. Van Nostrand Co., 1961.

12. Nelson, W., *Applied Life Data Analysis*, New York: John Wiley and Sons, 1982.

13. Ott, E. R., *Process Quality Control*, New York: McGraw-Hill Book Co., 1975.

14. Scherr, G. H., *The Journal of Irreproducible Results: Selected Papers*, 3rd ed., New York: Dorset Press, 1986.

15. Siegel, S., *Nonparametric Statistics for the Behavioral Sciences*, New York: McGraw-Hill Book Co., 1956.

16. Tobias, P. A., and Trindade, D. C., *Applied Reliability*, New York: Van Nostrand Reinhold Co., 1986.

17. U.S. Department of Defense, Military Standard 105D (Mil-Std-105D), *Sampling Procedures and Tables for Inspection by Attributes*, Washington, D.C.: U.S. Government Printing Office, 1963.

Journals

Biometrika, quarterly journal of Biometrika Trust, London, England.

Industrial Quality Control, former monthly journal of American Society for Quality Control, Milwaukee, Wisconsin.

Journal of Quality Technology, quarterly journal of American Society for Quality Control, Milwaukee, Wisconsin.

Quality Progress, monthly journal of American Society for Quality Control, Milwaukee, Wisconsin.

Technometrics, quarterly journal of American Society for Quality Control and American Statistical Association.